Cryptid Sea Monsters

CRYPTID
Sea Monsters

Kelly Milner Halls

Illustrated by
Rick C. Spears

little bigfoot
an imprint of sasquatch books
seattle, wa

Manufactured in China by Dream Color in March 2025

LITTLE BIGFOOT with colophon is a registered trademark of Blue Star Press, LLC

29 28 27 26 25   9 8 7 6 5 4 3 2 1

The authorized representative in the EU for product safety and compliance is Authorised Rep Compliance Ltd., Ground Floor, 71 Lower Baggot Street, Dublin D02 P593, Ireland. www.arccompliance.com

Editors: Christy Cox and Avalon Radys
Production editor: Peggy Gannon
Designer: Tony Ong

Library of Congress Cataloging-in-Publication Data
Names: Halls, Kelly Milner, 1957- author. | Spears, Rick, illustrator.
Title: Sea monsters : a field guide / Kelly Milner Halls ; illustrated by
   Rick C. Spears.
Description: Seattle, WA : Sasquatch Books, [2025] | Audience: Ages 7 |
Audience: Grades 2-3 | Summary: "Sea Monsters: A Field Guide profiles 50
different aquatic cryptid mysteries around the world.
Identifiers: LCCN 2024041320 | ISBN 9781632175168 (paperback)
Subjects: LCSH: Sea monsters--Juvenile literature.
Classification: LCC QL89.2.S4 H35 2025 | DDC 591.77--dc23/eng/20250218
LC record available at https://lccn.loc.gov/2024041320

ISBN: 978-1-63217-516-8

Sasquatch Books
1325 Fourth Avenue, Suite 1025
Seattle, WA 98101
SasquatchBooks.com

MIX
Paper | Supporting
responsible forestry
FSC
www.fsc.org   FSC® C188448

This book is dedicated to all the kids who loved *Cryptid Creatures*. It's also dedicated to Jhoskar, a young man I met at Warsaw Elementary in upstate New York. Promise kept, Jhoskar. I hope you love this one as much as you love *Cryptid Creatures*.

–KMH

Dedicated to Darien, Georgia—the home of Altie—and all other communities that have come to embrace their local marine cryptids. May your legends live forever!

–RS

CONTENTS

Sea Monster Extras

Appendices

INTRODUCTION

Marine animals have been observed since the beginning of human history. But sightings were not recorded in writing until the people of Mesopotamia—the land now known as the Middle East—created an alphabet and words to record human experiences.

Once the written word existed, water-based life forms were scientifically named and documented worldwide. Those animals were considered real.

When people learned to build ships capable of oceanic voyages, those descriptions expanded. Early maps of the seas featured illustrations of strange animals seen in the water. Remember, a full 96.5 percent of the world's geography exists under ocean waters, so the sightings were plentiful.

Creatures we know well today—such as whales and sea lions—were mysterious then. And clear observations were not always possible. So our familiar favorites were once seen as monsters.

Hundreds of marine animals have been scientifically confirmed—proven real. But others have remained a mystery. Fifty of those strange creatures are profiled in this book, thanks to careful research collected from reliable, journalistic sources.

Are all fifty real? It's hard to be sure—for now. Over time, some will turn out to be real, just as whales and sea lions turned out to be real. And some will be proven fictional. Most will remain unknown possibilities until new evidence comes to light. So consider that as you read about these water-bound cryptid creatures.

To guide you, we've created a reality rating. Will you agree with all of our ratings? Maybe not. But that's okay. At this point, your guess is as good as ours. In fact, we celebrate your theories, even if they disagree with ours. We all love a good mystery. So dig in and let us know what you think. And keep your eyes open when you're close to any body of water. You may be the one to prove these monsters really do exist.

Reality Ratings

Until definitive information is available, each cryptid will be given a reality rating.

★   Proven unreal

★★   Leaning toward unreal

★★★   Leaning toward real

★★★★   Proven real

Sea Monsters

Adult

ALKALI LAKE MONSTER

(AL-kuh-lie)

FIRST REPORTED	**1921**
LOCATION	**NEBRASKA, UNITED STATES**
CRYPTID TYPE	**LAKE MONSTER**
REALITY RATING	★ ★ ★

FACTOID: According to eyewitnesses, the Alkali Lake Monster has a foul smell you might detect even before you see it.

EYEWITNESS ACCOUNT: First reported in 1921, this lake monster is known by three different names—the Alkali Lake Monster, the Walgren Lake Monster, and *Giganticus brutervious.*

Six eyewitnesses stepped forward in 1922 to describe the creature. They said it was a twenty-foot-long reptilian Mermaid with scaly skin and razor-sharp teeth. Its roar was nearly deafening, and it had the power to project water ten to twelve feet in the air.

A 1922 *Hay Springs News* article said the witnesses were trustworthy so the story must be true, and monster mania spread. By 1923, new witnesses offered a new description—a forty-foot-long monster with a rhinoceros-like horn.

The Hay Springs Investigation Association was formed to collect more evidence that same year. But when the lake's owner demanded $4,000 for regular access to the shore, the association couldn't afford the fee, and the search was halted.

In 1962, writer Jack Reid wrote about the aquatic beast for *Outdoor Nebraska*. His description didn't match earlier reports. Reid said it was a three-hundred-foot-long dragon with a gaping mouth and a thunderous roar. When the monster wags its tail, Reid wrote, "the ground shakes."

Hay Springs Mayor Richard McKay even said the monster ate cattle and scared local farmers in 2017. Most citizens aren't sure what to think, but they celebrate every year with a parade and T-shirts featuring the creature's image. There is even an Alkali Lake Monster costume on display at the Hay Springs Heritage Center and Museum.

Local historian Mary Hansen says, "People saw something." So she keeps an Alkali Lake Monster scrapbook just in case the stories turn out to be true.

Alkali Lake
Monster skull

Baby

ALTAMAHA-HA

(al-ta-ma-HA-ha)

FIRST REPORTED	1500s
LOCATION	GEORGIA, UNITED STATES
CRYPTID TYPE	RIVER MONSTER
REALITY RATING	★ ★ ★ ★

FACTOID: In the sixteenth century, French explorer Jacques Le Moyne set out to draw every animal living in the territory now known as Georgia. One of his sketches appears to be the Altamaha-ha.

EYEWITNESS ACCOUNT: If the Altamaha-ha is hiding in Georgia, the Altamaha River is vast enough to keep its secrets. Nearly 140 miles long, most of the river flows far from populated areas until it meets the Atlantic Ocean near the city of Darien and St. Simons Island.

Armor-plated sturgeon fish up to twenty-three feet long thrive in the river's deepest waters. If the river has enough fish to feed a one-thousand-pound sturgeon, it could probably sustain a second large predator nicknamed Altie.

Adult

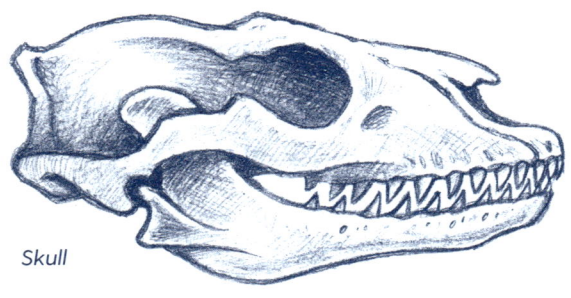
Skull

The Eastern Muscogee Creek, Yuchi, Shawnee, Apalachee, and Chickasaw peoples told stories of a river monster long before European explorers arrived. But written documentation came via the colonists.

On April 18, 1830, the *Savannah Georgian* featured a sighting by Captain Nathaniel Delano of the *Eagle*, a sailing ship, and five of his crew members. They claimed to have seen a seventy-foot aquatic monster with the girth of a rum barrel—thirty-six inches around its middle, big enough to hold fifty-three gallons of the alcoholic drink. It had a long neck and a head with a pointed snout.

Hundreds of similar sightings soon followed. Most described a gray-green-colored body with tire-tread ridges on top and a white belly. Large seal-like flippers propelled Altie through the water at great speeds. At the end of its tail was a diamond-shaped rudder. The witnesses say Altie does not swim with the side-to-side motion of a fish, but rather undulates top-to-bottom like a dolphin or a seal.

Hard evidence—body parts—was scarce until March of 2018 when Jeff Warren spotted a carcass on the Altamaha shore. He thought it was a dead seal at first, but when he got closer he saw flippers and a diamond-tipped tail. He took video and still photos and sent them to the *Savannah Morning News*. Sadly, the body had disappeared by the time reporters arrived.

An artist named Zardulu claims what Warren found was a fake monster she had made with a stuffed shark and paper-mache. Which claim is true? It's hard to know. Until another body appears and is examined by scientists, it will remain a river monster mystery.

Baby

BEAR LAKE MONSTER

FIRST REPORTED	1868
LOCATION	IDAHO/UTAH, UNITED STATES
CRYPTID TYPE	LAKE MONSTER
REALITY RATING	★ ★ ★

FACTOID: Indigenous peoples allegedly called this monster a "devil fish." The Shoshone, Ute, and Bannock tribes hunted near Bear Lake, but who coined the term isn't clear.

EYEWITNESS ACCOUNTS: When Mormon settlers discovered the deep blue waters of Bear Lake in 1863, they thought they'd found a heavenly pool. But the local Indigenous peoples quickly shared ominous warnings.

Within that vast pool, they said, was a serpent with a mouth big enough to swallow an adult whole. It had allegedly drowned adults swimming in the lake and snatched children from the shore as snacks.

The Mormon colonists soon agreed, and reports appeared in their hometown newspaper, Utah's *Deseret News*. In 1868, S. M. Johnson said he tried to help a drowning man, but ran when he saw the monster's head rise three feet out of the water.

Wagon train captain William Budge reported a similar sighting to the newspaper in May 1874. Then, in August 1881, Mormon church official George Q. Cannon saw the monster as he walked the shore of the lake.

Most agreed the Bear Lake Monster had a thick brown body about forty feet long. Its head was large, like that of a walrus without tusks. It had two oversized eyes about a foot apart and ears on each side of its head. Some said it had multiple legs, about eighteen inches long, but had trouble walking on land. Once it hit the water, it could swim up to sixty miles per hour.

In time, the colorful stories made their way to Brigham Young, the leader of the Church of Jesus Christ of Latter-day Saints—the Mormons. Young traveled to the lake to investigate, according to the *Deseret News*. He came to believe the witnesses were trying to tell the truth but were confused. He concluded it was just a large fish.

Adult

Phineas W. Cook decided to catch that fish. He planned to bait a giant hook with a slab of lamb meat, attach it to three hundred feet of rope, anchor the land end of the rope on a tree, and set the hook when the time was right. Cook asked Brigham Young to pay for the rope, and he did, but the plan failed. The bait was removed from the hook while it was submerged, but the creature escaped.

Skull

Decades later, writer Alan Edwards floated a new theory in the *Deseret News*. In 2003, he said the Bear Lake Monster and the Loch Ness Monster were the same animal. He claimed there was an underground tunnel from Idaho to Scotland that made it possible. Edwards's theory is highly unlikely. And so, the search continues.

Baby

Making the Bear Lake Monster Movie

Utah-based filmmaker Brandon Smith isn't sure if the Bear Lake Monster is real or not. But that hasn't stopped him from writing, directing, and producing a movie about it. Titled *The Legendary Bear Lake Monster*, it follows three scuba-diving teens as they search for evidence to prove the creature is more than a tall tale.

While he admits the monster may not be real, Smith is enchanted by the possibilities. "There is magic in believing," he says.

Smith's film is an action-adventure with the suspense of *Jaws* and the heart of *The Goonies*. He hopes it will entertain viewers of all ages. A trailer is available on YouTube (search "Brandon Smith" and "Bear Lake Monster movie" to find it). Financing for the film came from an Indiegogo fundraising page dedicated to keeping the Bear Lake Monster legend alive for generations to come.

Adult

BEISHT KIONE DHOO/ BEAST OF BLACK HEAD

(by-ISHT KEY-on Dhu)

FIRST REPORTED	UNKNOWN
LOCATION	ISLE OF MAN, OFF THE COAST OF ENGLAND
CRYPTID TYPE	SEA MONSTER
REALITY RATING	★ ★

FACTOID: The name Beisht Kione Dhoo originated in the Manx Gaelic language. It means the "Beast of Black Head."

EYEWITNESS ACCOUNTS: The ancient fishermen of the British Isles relied on the sea to make a living. But going to work could be dangerous thanks to thundering waves, the weather, and the business of sea monsters.

One creature feared by all was the Beisht Kione Dhoo. According to ancient stories, the beast made its home in sea caves of a rocky outcropping called Black Head, on the southern tip of the Isle of Man.

Skull

When it emerged from its rocky sanctuary, Beisht Kione Dhoo gave people a lot to talk about. It allegedly had the hairless head of a large horse. Its mouth was lined with dagger-like teeth, and its huge eel-like body dwarfed other marine animals.

Baby

Because it was such a deep black color, it was hard to detect in time to protect a ship or crew. In fact, once the shine of its yellow eyes was spotted, it was too late to escape.

Beisht Kione Dhoo's attacks were powerful and swift. It could attack with its teeth or wrap its powerful body around its target and squeeze with crushing coils. Some witnesses said it could even leap out of the water to eat children who wandered too close. Beisht Kione Dhoo could allegedly produce offspring without a mate. As it died, its young ate their way out of the parent's body.

Legend says the first Beisht Kione Dhoo began its life as a man who discovered a pirate's hidden treasure. The pirates killed him to keep the treasure's location a secret and hid his body in a cave. In time, the dead man was transformed into a monster with a ferocious roar and a taste for blood.

Only one thing could calm Beisht Kione Dhoo—an alcoholic drink called rum. When fishermen set sail, they tossed a glass of rum overboard as an offering. In return, the sea monster would help them catch a bounty of fish to bring home. So carry a mug of rum if ever you visit these waters.

BESSIE

(BESS-ee)

FIRST REPORTED	1793
LOCATION	THE UNITED STATES AND CANADA
CRYPTID TYPE	LAKE MONSTER
REALITY RATING	★ ★ ★

FACTOID: A man reportedly found four eggs on the banks of Lake Erie in 1898. When they hatched in his kitchen sink, he believed they were two-foot-long baby Bessies.

EYEWITNESS ACCOUNTS: In 1793, a duck hunter rowed his boat across Lake Erie. When he fired his shotgun, the sound woke the Lake Erie monster. According to his report, the enormous sixteen-foot serpent broke the calm of the lake, frightened by the gunfire. But it vanished as quickly as it had appeared before the hunter could get a closer look. That mystery monster was soon nicknamed Bessie.

The *Cleveland Plain Dealer* covered another sighting on May 21, 1892. A fisherman spotted the creature near Oak Harbor, Ohio. He described a twenty-five-foot-long monster

Skull

nearly two feet wide. It was black with brown spots and had flippers and a flat, wide head.

Employees of the Ohio Salt Company claimed to see baby Bessies on the surface of Lake Erie in 1909. They played together for a few minutes and then swam toward Cleveland and submerged.

Sightings have been numerous. So have monster nicknames. It's been called the Great Snake of Lake Erie, the Lake Erie Monster, and South Bay Bessie.

Some say Bessie has a dog-like head and large fins. Some say it has bright red eyes and humps that are visible above the water. Some say it has arms that look almost human. Some say it is not black, but rather gray, copper, or silver.

Iroquois tradition says the creature is the Oniare, the ancient water spirit of the Great Lakes. Oniare translates to "snake." Oniare would spit fire and poison to sink boats and then eat the crew. Is Oniare the same creature seen more recently in the lake? Perhaps.

Adult

In the 1980s, the *Port Clinton Beacon* held a contest to name the mysterious beast. "South Bay Besse" was the winner. It was inspired by the Davis–Besse nuclear power plant in Toledo, Ohio. It evolved into "Bessie" almost immediately.

In 1990, the *Daily Kent Stater* published the story of Harold Bricker and his family. They saw a thirty-five-foot-long black serpent only one thousand feet from their boat. It was fast enough to keep up with their boat before it submerged and swam away.

Stories are not necessarily proof. But if Bessie is out there, the people who live and work near Lake Erie are sure to notice.

Baby

Monster Babies with Bite

Bessie the Lake Erie Monster has been spotted quite often. But in August of 2001, a frightening new monster made itself known. Nicknamed "Chomper," the sharp-toothed predator attacked three different swimmers in the course of twenty-four hours.

One of the victims was Brenda McCormack. When she slipped into the waters of Lake Erie at sunset, she felt something bite her right calf. Startled and in pain, McCormack made her way to the beach to discover a series of circular puncture wounds in the shape of a jaw in her right calf.

An unidentified man and his son were the next victims. The day after the McCormack event, they felt a series of bites as they waded through the waves of Lake Erie. They rushed to a medical facility to have the injuries checked, and the son required hospitalization.

Dr. Harold Hynscht, who treated all three victims and who has been known to scuba dive in Lake Erie, says he has a mystery on his hands. He confirms that the bites were not minor injuries. They measured about six inches around, the top teeth on one side of the circle and the bottom teeth on the other. "That's a big, honking fish," Dr. Hynscht said.

He ruled out gobies, lamprey eels, snapping turtles, walleye, muskellunge-type fish, and piranhas early in the examinations. But he says the bowfin is a strong possibility. When they spawn, they are aggressive for up to nine weeks as they guard their eggs.

Nicknamed "Chompers," they could cause more trouble in the years to come. If Bessie has babies of her own, let's hope they scare the Chompers away.

CADBOROSAURUS

(CAD-bor-oh-sahr-us)

FIRST REPORTED	**1791**
LOCATION	**BRITISH COLUMBIA, CANADA**
CRYPTID TYPE	**SEA MONSTER**
REALITY RATING	★ ★ ★ ★

FACTOID: Pacific Northwestern Indigenous people captured the likeness of *Cadborosaurus* in petroglyphs.

EYEWITNESS ACCOUNTS: Centuries ago, stories were told of a giant serpent migrating from the northern waters of Alaska to Canada's Cadboro Bay as the waters warmed.

Today, that serpent is called Cadborosaurus—Caddy for short. Most agree Caddies are forty to fifty feet long as adults and ten to fifteen feet long as babies. They have lean, muscular bodies covered in visible green scales. Their heads resemble those of hairless horses or camels, and they propel their bodies through the ocean with powerful flippers.

In 1937, a black and white photo of what is known as the Naden Harbor carcass surfaced. One team of scientists and

Baby

fishermen said it was the image of a baby Cadborosaurus cut from the stomach of a dead whale. When others said it was a whale calf not yet born, the Caddy believers reminded them it was removed from the stomach, not the uterus, of the whale.

For a time, that photograph was the strongest evidence for the sea beast. But Captain William Hagelund added to the story in 1968 when he captured a baby Caddy in a bucket.

He first spotted the baby swimming in the bay. He quietly observed as it swam closer to his boat and turned its curious head from side to side to get a better look at its surroundings. Hagelund scooped the baby up in a large bucket of sea water. It appeared to be about eighteen inches long with a body as big around as the captain's thumb. He noticed tiny claws on its feet as it tried to escape the bucket.

Hagelund took the bucket home and checked on the baby throughout the night. When he noticed the baby was exhausted, he feared it might drown, so he took it to the water's edge and let it go.

When Hagelund prepared to write a book on his life as a sea captain, he discovered the Naden Harbour carcass photo and realized its description matched the creature he'd once kept in a bucket.

Decades later, oceanographer Paul LeBlond and zoologist Edward Bousfield continued the search for Cadborosaurus evidence. Together, they compiled hundreds of eyewitness reports, photographs, and film clips to determine whether or not it could be declared a scientific species rather than a rumor.

Adult

Once they amassed enough evidence, they proposed giving Caddy a scientific name—*Cadborosaurus willsi*. They were denied, but they published huge portions of their research in 1995 in a book called *Cadborosaurus: Survivor from the Deep*. Their scientific approach was admired but also discounted.

LeBlond and Bousfield were not cryptid hunters. They were scientists with real credentials. If Cadborosaurus was a hoax, wouldn't they have been the first to recognize it? Before his death, LeBlond admitted it's a hard theory to prove. "Caddy sightings are incredibly rare," he said, "and until an actual specimen is found, the existence remains only a possibility."

Skull

Adult

CAPE ANN SEA SERPENT

FIRST REPORTED	1639
LOCATION	MASSACHUSETTS, UNITED STATES
CRYPTID TYPE	SEA MONSTER
REALITY RATING	★ ★ ★ ★

FACTOID: Some experts say this sea serpent was actually a narwhal, with its signature sword-like tusk.

EYEWITNESS ACCOUNTS: In 1639, a man named John Jossellyn heard and shared a story that may have launched the legend of the Cape Ann Sea Serpent.

An unnamed witness told Jossellyn, "A sea serpent lay coiled like a cable upon the rock at Cape Ann." As a boat carried two Englishmen and two Indigenous people toward the rock, one of the Englishmen aimed a rifle at it, but the Indigenous men stopped him. If he missed, they said, the monster would destroy them.

Similar stories soon followed. So the Linnaean Society of New England, a group devoted to Boston's natural history, put the evidence in a booklet in 1817. The society hoped to give the creature a scientific name—*Scoliophis atlanticus*—and make it official. They were denied.

Forty-two years later, General David Humphreys, commander of a ship called the *British Banner*, spotted the sea serpent once again. He saw a sixty-foot-long creature with a twelve-inch spear protruding from its forehead. Its eyes, he said, were bright and glistening. It sped by so quickly, Humphrey's crew was unable to kill it.

Baby

The fact that the witnesses saw the marine animal undulating—swimming with an up-and-down motion rather than the usual side-to-side motion like a fish—was regularly reported.

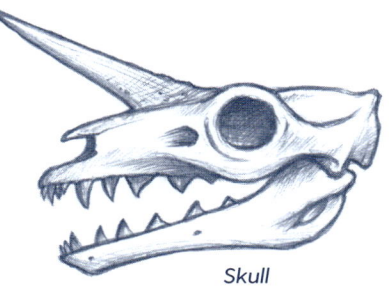

Skull

Sea Captain Gresham Bennett added to the reports, saying, "I saw a serpent of enormous magnitude floating on the water. Its head rose about seven feet above the surface. The color of the animal in all its visible parts was black. The skin appeared smooth and free from scales. Its head was about as long as that of a horse but was the perfect head of a serpent."

Bennet swore he watched the creature for eight minutes as it swam beside his boat. He said it had humps the size of a large barrel.

Sightings dwindled with the creation of motorized boats and ships, and some feared the sea monster was frightened away by the ferocious roar of the motors. But on Sunday, November 15, 1970, new evidence was revealed.

A sea monster body allegedly washed up on the shore near Boston. In a three-day period, more than ten thousand people came to see the decomposing carcass. People chopped off pieces of the body as souvenirs before scientists from the New England Aquarium came to investigate. The mutilation made it impossible to identify what the animal actually was.

Was it the Cape Ann Sea Monster? Or did another lifeless body mislead the Massachusetts witnesses? Those questions remain unanswered.

CHAMP/GITASKOG

(GIT-a-scahg)

FIRST REPORTED	1609
LOCATION	**VERMONT/NEW YORK, UNITED STATES AND QUEBEC, CANADA**
CRYPTID TYPE	**LAKE MONSTER**
REALITY RATING	★ ★ ★ ★

FACTOID: P. T. Barnum—the "Greatest Showman" depicted in the 2017 Hollywood musical—offered a $50,000 reward for the capture of Champ, dead or alive.

EYEWITNESS ACCOUNTS: In 1609, French explorer Samuel de Champlain set out to map the wilderness of Vermont; New York; and Quebec. The Indigenous Abenaki warned him of a lake monster they called Gitaskog. De Champlain never saw the creature, but he sketched it in his journal based on what the Abenaki described.

Today, that monster of Lake Champlain, named for the explorer, is called Champ. More than three hundred sightings have been documented. Another three hundred have been shared but not published.

Adult

Lake Champlain is a huge body of water—125 miles long and up to four hundred feet deep in some locations—but many of the sightings have come from Bulwagga Bay, near the southern end of the lake on the New York side.

Captain Crum, who sailed a barge on the lake, made his Champ report from Bulwagga Bay in the *Plattsburgh Republican* on July 22, 1819. He said the monster was about 180 feet long and black in color, with a head that resembled that of a seahorse. He said it was close enough to see clearly when it reared fifteen feet out of the water. He described a mouth with "three teeth, eyes the color of a peeled onion, a white star on its forehead, and a belt of red around the neck."

The *New York Times* covered a sighting in 1873 after a railroad work crew said they'd seen a supersized sea serpent with silver scales glistening in the sun. Clinton County Sheriff Nathan Mooney saw it that same year. And tourists aboard the steamship *W. B. Eddy* feared their boat would sink when it collided with the mysterious beast.

In July of 1977, Sandra Mansi secured the best evidence of Champ ever revealed—an iconic color photograph taken with a Kodak Instamatic camera. As she picnicked with her family, Mansi saw a giant head rise up out of the water near where her children were splashing in the lake. It turned to face her, and when it did, Mansi saw water drip out of its mouth. She snapped a

Baby

picture then fled with her children, afraid the monster might be dangerous.

Champ/Gitaskog skull

When she developed the film, Mansi knew she'd captured proof, but she didn't share it until 1981 for fear of public reaction. When the *New York Times* published Mansi's photo, some said it was indisputable proof of Champ's existence.

When skeptics asked to see Mansi's negative of the photographic image, she had to admit she'd thrown it away. Unbelievers said that proved the photo was fake. But photography experts that analyzed the picture said it was legit.

Mansi got more support in 1984, when eighty passengers on the ship called the *Ethan Allen* saw Champ swimming alongside the boat. They reported three to five humps undulating in the water before it submerged as a motorized speed boat approached.

Two swimmers in Vermont said baby Champs swam with them in Button Bay State Park in 1993. And in 2005, fishermen Pete Bodette and Dick Affolter captured video of what they say was Champ near the Ausable River.

Sonar and sound evidence suggests that something large and unidentified does seem to live in the lake, but DNA has yet to be as conclusive.

We'll need more evidence to be sure Champ is real. But if you ever search for Champ in the wild, don't bring a gun. Vermont and New York State have passed legislation to protect the mysterious beast.

Adult

CHESSIE

(CHESS-ee)

FIRST REPORTED	1934
LOCATION	MARYLAND, UNITED STATES
CRYPTID TYPE	LAKE MONSTER
REALITY RATING	★ ★ ★ ★

FACTOID: Seventy-eight different reports were recorded between 1982 and 1988.

EYEWITNESS ACCOUNTS: Fishing fans have claimed to have seen Chessie since 1934, when Francis Klarrman and Edward Ward first reported seeing the all-black sea monster. They watched the twelve-foot-long creature "coming up for air." As it swam past their boat, the sea serpent turned its head to look at Klarrman and Ward, as if it was curious. They said the head was "about as big as a football," and shaped like a horse's head. They wondered if it might be a dinosaur that had escaped extinction.

Other fishermen made similar reports, but two sightings from the air really expanded the reach of the legend.

In 1943, an unnamed military helicopter pilot flew over the bay. His crew saw something "reptilian and unknown." Once again, the theory of a dinosaur or marine reptile from prehistoric origins was one theory, but they did not report additional descriptive details.

When a second helicopter pilot named Walter Myers spotted Chessie in 1963, he documented what he'd seen in a letter to his state senator. "I assure you that Chessie exists," he wrote, and he hoped the state would investigate further. Although the letter doesn't prove the sea monster is real, it does suggest the name took root in the early 1960s.

In 1977, Greg Hupka saw Chessie and made an attempt to capture photographic evidence. He claimed to capture a shot of Chessie raising its head above the water. Skeptics said it was the head of a turtle, but Hupka stood firm. "No way it could have been a turtle," he countered. It was Chessie.

Most eyewitnesses agree Chessie has no flippers or limbs. It resembles a huge eel or sea snake rather than a plesiosaur. Most say it is black, brown, or bluish gray and measures between twelve and twenty feet long.

On September 7, 1980, Bob White Sr. saw "a large, strange, and awesome thing that was no manatee." According to White, the creature raised its head four feet out of the water just ten feet from his boat. As he moved past the beast, he was sure it was like no animal he'd ever seen in the bay before.

Video evidence was captured in 1982. "There were five or six of us here," said Bob Frew. "We all saw it. It went up in the rocks to a place called Cloverfields, where a commercial artist sketched it. It was exactly what we saw on the film." Frew shared the video with WJZ-TV, and other witnesses soon came forward.

Clyde and Carol Taylor saw the sea creature just a few miles from where Frew took his video. Other eyewitnesses agreed, including FBI and CIA agents, navy and coast guard officers, and commercial fishermen.

In 1983, John Hoffman and his friend Ruben Ribaya were fishing when they saw the most vivid account of a Chessie sighting yet recorded. The water was smooth and calm, like a mirror. When they saw the creature, they thought it was a floating telephone pole. But as they got closer, they realized it was not floating, it was swimming with the side-to-side motion of a snake and leaving an eight-inch wake.

Baby

When sightings slowed after 1983, newspapers claimed "Chessie is dead," but witness Chris Gardner countered the claim in 2014. He saw Chessie five feet from the passenger seat of his car. "It was twenty feet long," he wrote. "Its head and tail breached the surface."

Gardner confirmed it looked glossy and black. Its head seemed to be hooded, like a cobra. "I didn't think to take a picture with my phone, which was in my pocket," he said. But when he reported it to the Maryland Department of Natural Resources, he said, "They didn't laugh."

Skull

Chessie, a Multicolored Sea Monster

In 1986, the Maryland branch of the US Fish and Wildlife Service created a promotional coloring book with a very special star—the monster of Chesapeake Bay. Why Chessie? The monster's fans want to keep it safe. And the best way to keep Chessie safe is by protecting its habitat, Chesapeake Bay.

Pollution is a danger to all waterways and marine creatures on earth. Millions of tons of trash—plastic bottles, candy wrappers, and even lawn chairs and tires— are dumped in the bay each year. Every scrap endangers life, including Chessie.

As an environmental assistant to the Maryland Department of the US Fish and Wildlife Service, Chessie encouraged kids to learn more about Maryland by coloring in the book, *Chessie: A Chesapeake Bay Story*. If young people learn to be responsible citizens as they color, that's a win-win.

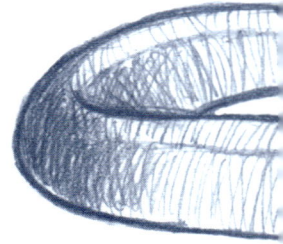

CRESSIE

(CRESS-ee)

FIRST REPORTED	1950s
LOCATION	NEWFOUNDLAND, CANADA
CRYPTID TYPE	LAKE MONSTER
REALITY RATING	★ ★ ★ ★

FACTOID: Cressie was named as a tribute to the Loch Ness Monster, Nessie.

EYEWITNESS ACCOUNT: Rumors of a monster in Crescent Lake, Newfoundland, have been whispered since the 1950s. But it was Fred Parsons who really put Cressie on the map.

In 1991, the former teacher was driving on the road beside Lake Crescent when he spotted something odd. "It seemed to be just lying on the surface," Parsons said. It looked like a giant eel—brown and twenty feet long. When Parsons shared

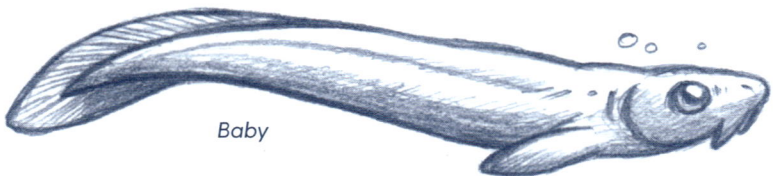

Baby

Adult

his experience with neighbors, dozens admitted they'd seen Cressie too.

Scientists say eels thrive in Canadian waters. And old trees from the region's logging era litter the floor of the lake. They sometimes float up to the surface of the water, appearing to be otherworldly creatures as they drift with the flow. Could that be what Parsons and the others saw?

To add to the mystery, odd holes sometimes appear on the surface of frozen Lake Crescent in the winter. Ice fishermen sometimes cut holes for their hooks and poles, but these had not been cut from above. They'd been punched from beneath, according to Memorial University folklore expert, Nicole Penney.

Could the holes have been made by Cressie? Perhaps. But there is another possibility.

Scuba divers used the winter holes to search for a missing pilot in the 1980s. Once they slipped into the icy water, they were surrounded by giant eels. "Their bodies were as thick as a man's thigh," Penney said. The eels attacked with gusto, but the divers were lucky enough to escape without serious injuries.

Were the winter eels hungry baby Cressies? Believers think it's possible. But more evidence is required to know for sure.

Skull

EDIZGIGANTUS

(ed-IZ-gig-ant-us)

FIRST REPORTED	**1933**
LOCATION	**WASHINGTON, UNITED STATES**
CRYPTID TYPE	**SEA MONSTER**
REALITY RATING	★ ★

FACTOID: In 1934, painter Thomas Henry Guptill created a portrait of the *Edizgigantus* sea monster based on eyewitness reports.

EYEWITNESS ACCOUNT: On October 11, 1933, the *Victoria Daily Times* ran a story about a giant sea monster swimming through the coastal waters of British Columbia, Canada. Most believed it was a publicity stunt created by the newspaper's editor, Archie Wills. The Loch Ness Monster was a hit in Scotland, and he may have thought a Pacific Northwestern monster would sell papers too.

Canada called it the Cadborosaurus. But Washington State named it Edizgigantus—the monster Guptill

Adult

painted—in honor of the Ediz Hook Lighthouse in Port Angeles, Washington.

An eyewitness letter published more recently in the *Victoria Daily Times* came from a man driving along Lake Washington by Seward Park in Seattle, Washington, which is over 150 miles from Victoria. It was about 5:30 p.m. and it was raining. The man had seen otters and beavers playing on rainy days before, so he was hoping to see them again.

As he came to the entrance of Andrews Bay, he spotted something grayish blue in the water causing ripples on the smooth surface. It appeared to have three humps and a serpentine body. Could it be Edizgigantus?

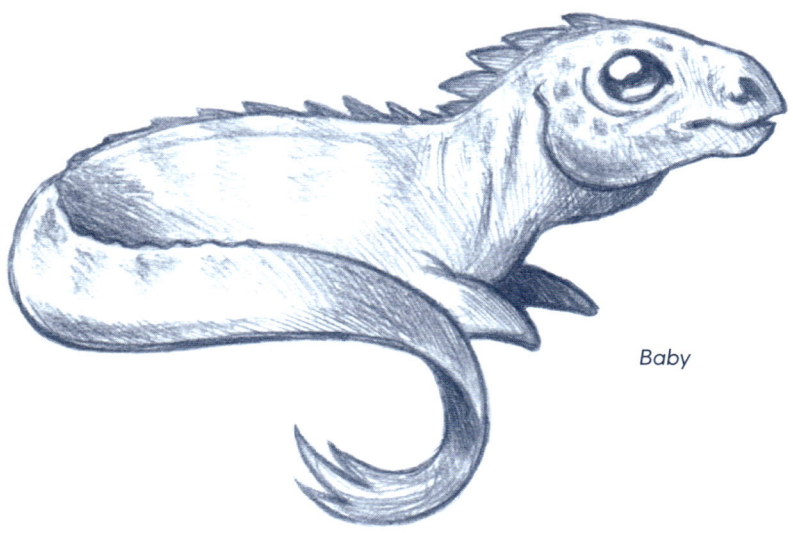

Skull

His heart was pounding as he got closer to the beast. Then he had to laugh. It was not Edizgigantus. It was a chain of three ordinary rowboats. The man was disappointed at first. Then he remembered the Cadborosaurus in the Salish Sea, and his hope was revived. He might still see a sea monster someday.

Some Pacific Northwesterners believe Guptill's painting captured Cadborosaurus. Some say Edizgigantus is a sea monster of a different kind. Without further evidence, the mystery will remain unsolved.

Baby

FLATHEAD LAKE MONSTER/FLESSIE

(FLESS-ee)

FIRST REPORTED	1889
LOCATION	MONTANA, UNITED STATES
CRYPTID TYPE	LAKE MONSTER
REALITY RATING	★ ★ ★ ★

FACTOID: Though people call it a "monster," most eyewitnesses say this creature isn't really scary. It's just big.

EYEWITNESS ACCOUNT: The Flathead Lake Monster is a massive creature with steel-black eyes that was first spotted in 1889. More than one hundred passengers aboard the steamboat *US Grant* and the captain, James C. Kerr, saw something big headed right toward the boat.

At first, they thought it was a floating log. But when it was alongside the boat, they realized it was a twenty-foot-long serpent or eel splashing in the wake. One passenger pulled out a rifle, hoping to kill the creature, but he missed and it escaped.

More witnesses soon followed. Laney Hanzel, a biologist with the Montana Department of Fish, Wildlife and Parks, started interviewing them in 1990. When he retired in 1993, he began to seek out more stories in his free time.

Hanzel has never seen the lake monster himself, but he does believe most people are telling the truth. And he has seen enough fishing nets with unexplained holes gnawed through them to keep an open mind.

In thirty years, Hanzel collected 109 stories, and eighty of them are nearly identical in details. They all describe a serpent-like creature that's thirty feet long with black eyes. It undulates through the lake waters like a marine mammal. Those similarities keep Hanzel wondering.

Adult

Credibility of the witnesses also keeps the monster theories alive. In 2005, Lake County judge Jim Manley and his wife Julia were boating on the lake when their engine failed. They used their cell phone to call for assistance and settled in to wait.

In the quiet, Manley heard a strange noise—a loud splashing from the opposite side of his boat. When he turned to investigate, he was astonished. Manley had heard the Flathead Lake Monster stories for years, but now he was seeing it with his own eyes, just a few feet from the boat.

To this day, Manley and Julia still insist the Flathead Lake Monster is not a legend—it's real.

Tami Avison had her own sighting while she was staying with her grandmother as a child. She watched four or five

Flathead Lake Monster babies playing in the bay just beyond her grandmother's window. When Avison asked what they were, her grandmother said, "Oh, those are our lake monsters."

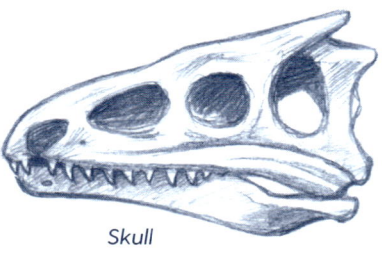

Skull

Hanzel believes he could double the number of his reports if everyone who's seen the monster would admit it. But most remain silent to avoid being called crazy.

Manley understands that concern. One person asked him if he saw ghosts too, after he shared his creature story. But he stands by his testimony. He believes the creature might be shy, avoiding human beings whenever possible.

Flathead Lake Monster pizzas, sodas, murals, and statues are scattered throughout lakeside communities. Few people are afraid of their mysterious neighbor. It's not dangerous, they say. And they hope to keep it safe for generations to come.

Baby

GHOSTLY EEL

FIRST REPORTED	**2022**
LOCATION	**SOUTH AFRICA**
CRYPTID TYPE	**SEA MONSTER**
REALITY RATING	★ ★ ★ ★

Tooth

FACTOID: Ghost eels—also called white ribbon eels, blue ribbon eels, and black ribbon eels—are part of the *Rhinomuraena* group of marine creatures. This new cryptid may be a member of that family.

EYEWITNESS ACCOUNTS: When thirty-seven-year-old diver Amy Wainman slipped into the waters of South Africa's Western Cape, she never imagined she'd witness something ghostly. But she got it on video. She filmed a transparent sea creature nearly five feet long.

At first, she thought it was floating plastic—garbage tossed into the ocean. When it started to swim, Wainman said it was "like a dancing, clear ribbon." She called it "magical" and said it was like nothing she'd ever seen before.

Baby

Experts in marine biology are not entirely sure what Wainman captured on video, but they are willing to offer a few theories.

Could it be a species of jellyfish? Kevin Kocot, an invertebrate zoology professor at the University of Alabama, thinks so. He believes it has a mouth at the middle of its body, something common to jellies. Better known as a Venusgirdle, it eats plankton, tiny fish, and crustaceans. Kocot believes this jellyfish has a mouthful of food in Wainman's video. They are safe to touch, Kocot says, but poisonous if eaten by human beings.

An eel is another possibility. Retired marine science professor Bradley Stevens believes it was a young eel that had just finished its first migration to South African waters. Soon, Stevens says, it will look more like other eels common to the region.

Wainman says it is one of a kind, regardless of what it turns out to be. And she hopes to one day see it again.

For now, the ghostly visitor does exist and has captured the imaginations of animal experts worldwide. What species it represents remains a mystery—for now.

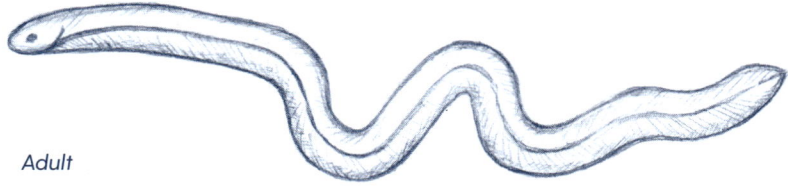

Adult

GONAKADET/WASGO/ SEA WOLF

(gon-ce-AH-kah-det)

FIRST REPORTED	7,000 BC
LOCATION	ALASKA, UNITED STATES AND BRITISH COLUMBIA, CANADA
CRYPTID TYPE	SEA MONSTER
REALITY RATING	★ ★

FACTOID: The Canadian Museum of History has a Sea Wolf carving in their collection—a tribute to this prehistoric cryptid.

EYEWITNESS ACCOUNT: The Gonakadet is also known as the Sea Wolf in Tlingit tradition. The Haida people call it the Wasgo. The same legendary cryptid living in the Pacific Northwest and Alaska is known by all three names.

Ancient stories tell of how the mysterious animal impacted tribal history. When a daughter married a man her mother didn't like, the young man would try to win her approval. To his amazement, the young husband captured and killed a

Adult

Gonakadet—a legendary creature with the head of a wolf and the body of a killer whale.

He skins the lifeless beast and wears its pelt on his shoulders to hide his identity. And wearing the skin, he absorbs the Gonakadet's magical powers. Food has been scarce in his village, but the powers make it possible for him to catch salmon and seals in abundance.

He places the riches on his wife's mother's doorstep, hoping to win her love. But she refuses to give him credit. Instead, she claims she has magical powers and gathered the feast on her own.

Heartbroken, the young man staggers to the shore and collapses. When the selfish mother discovers his broken body and soul, she is ashamed and dies, lying beside him.

Some say the story was only a warning: do not take credit for the good deeds of others. Others say the story is true and Gonakadets still swim the northwestern waters, hoping to escape capture.

One other explanation is the strong swimming skills of the gray wolves that claim the Pacific Northwest as habitat. As the wolves swim from island to island, people sometimes mistake them for legendary creatures of cryptozoology.

Which explanation is the truth? It's impossible to say. The Gonakadet is carved in totem poles. It's woven into baskets. It's painted on ancient petroglyphs, and it's even worn as tattoos on people who admire its strength and promise.

So keep your eyes on those ocean waves to gather more evidence of your own.

Skull

Baby

Adult

HIPPOCAMP

(HIP-po-camp)

Seahorse

FIRST REPORTED	4,000 BC
LOCATION	GREECE
CRYPTID TYPE	SEA MONSTER
REALITY RATING	★

FACTOID: NASA named the planet Neptune's moon Hippocamp after this fictional cryptid.

EYEWITNESS ACCOUNT: As much as we would like to believe this half-horse, half-fish creature could be real, it was a creation of Greek mythology more than six thousand years ago.

In the stories, a Hippocamp had the head and forelegs of a powerful horse, a mane made of shark fins, and the green-scaled tail of a fish. According to ancient Greek stories, the Hippocamp was the adult form of the tiny sea horses we still see in oceans today. They were so powerful, even against the ocean's raging waters, that Poseidon, the god of the sea, drove a chariot pulled by four of the glorious creatures. Even lesser magical sea creatures such as nymphs rode Hippocamps through the waves.

Ancient Greek literature claimed the Hippocamp could leave the sea and shake off the salt water to reveal the body of an earth-bound horse with a gleaming golden mane. When it returned to the waves, the Hippocamp's fish tail and scales returned as its back legs mysteriously vanished.

Hippocamps appear on many Greek works of art called mosaics. Created one tiny tile at a time, the images took shape thanks to shifting colors.

Many other fish-tailed creatures appeared in Greek mythology:

- Leokampos, a fish-tailed lion
- Taurokampos, a fish-tailed bull
- Pardalokampos, a fish-tailed leopard
- Aigikampos, a fish-tailed goat

While the creature of Greek mythology exists only in our imaginations, the distant moon called Hippocamp is real and orbits Neptune, about three billion miles from Earth.

Skull

Holy Beasts

Are sea monsters mentioned in the Christian book known as the Bible or the Muslim Quran? The answer is yes.

In the Bible's Psalms 74:12–14, just as God is creating the seas, the land, the sun, the moon, and the stars, he is also destroying sea monsters.

"You divided the sea by your might," the passage says. "You broke the heads of the sea monsters on the water. You crushed the heads of Leviathan."

In the Quran, a large fish swallows the prophet Yunus after he is thrown overboard during a storm. The fish is told to protect the holy man so he'll have time to learn patience. The Bible says God shared the meat of the sea monsters with Moses and his followers in the wilderness, as the former Egyptian slaves searched for a place to live free.

In Isaiah 27:1 and Job 26:13, the Bible describes the sea beast Leviathan as a "twisting serpent."

One work of Muslim fiction tells of a giant fish that holds the weight of the entire Earth. "It is so immense," the story says, "all the seas of the world placed in one of the fish's nostrils would be like a mustard seed laid in the desert."

Some say the sea monsters aren't meant to be taken literally. They are symbols of the enemies of God. Others believe the beasts once lived in the world's oceans. Either way, sea monsters are mentioned not only in the Bible and the Quran but also in the Talmud and the Bhagavad Gita!

Adult

ILIAMNA LAKE MONSTER

(ILL-ee-am-nah)

FIRST REPORTED	**4,000 BCE**
LOCATION	**ALASKA, UNITED STATES**
CRYPTID TYPE	**LAKE MONSTER**
REALITY RATING	★ ★ ★ ★

FACTOID: In 2016, an anonymous witness shared a photo of the Iliamna Lake Monster—a reptile with black skin, yellow markings, and yellow eyes.

EYEWITNESS ACCOUNT: Alaska's Iliamna Lake is expansive. It is seventy-seven miles long and twenty-two miles wide, and it is connected to the Bering Sea by a river that allows oceangoing animals such as seals, sea lions, and beluga whales to visit the lake. Fishermen catch salmon and rainbow trout in Iliamna Lake too. But none have hooked the legendary Iliamna Lake Monster—so far.

First described by Alaska's Indigenous peoples, this monster is reportedly fifteen to twenty feet long with a broad head and coloration that goes from silver to dark brown along its body. It swims swiftly, its tail moving from side to side like a fish. Some say it has a line of sharp predatory teeth and rows of fins along its back.

Bush pilot Babe Alyesworth and his friend Bill Hammersley spotted a ten-foot fish from the air in 1942 that might have been a small Iliamna Lake Monster. Larry Ross saw a twenty-foot fish from his plane in 1945.

In 1957, twelve-year-old Agnes was skipping rocks along the lakeshore when she noticed the water beginning to churn a couple hundred yards away. A creature with the head of a lion and the body of a serpent was moving toward the shore. Agnes ran home and told her grandmother, who told Agnes to stay away from the lake because the creatures fed in the spring and fall.

Chuck Crapuchettes saw the monster in 1967 from the air. He landed his float plane and baited a massive tuna hook with a huge slab of caribou meat. He fixed the hook to a stainless steel cable and attached the line to one leg of his plane. Once he dropped the hook into the water, he waited for the creature to swallow the hook.

The mystery monster did not disappoint. When it took the bait, the creature jolted the plane with such force,

Some believe this monster was a sturgeon.

Some believe it was a Pacific sleeper shark.

Crapuchettes nearly fell into the water. He felt his float plane move, trailing the beast. He feared his plane would be pulled underwater, so he cut the cable, and the giant fish disappeared.

In 2020, Yup'ik sisters Christina and Alexanna Salmon saw what the pilot had seen, but from a safer distance. They had heard stories of the Iliamna Lake Monster for decades. A mural in their school warned that the monster was attracted to red, so boats should be painted blue.

When the sisters saw the monster, they could barely believe their eyes. At first, they thought it was a whale. But it was too big and black to be the beluga whales that usually swam the waters.

When they remembered the old stories, they were convinced they'd seen the Iliamna Lake Monster, but they were not afraid. They saw the experience as a gift—confirmation that old stories still mattered.

Outsiders believe the monster is a Pacific sleeper shark that swam up the Kvichak River to the lake from the Pacific Ocean. But eyewitnesses disagree. They insist it's the Iliamna Lake Monster, and they hope to keep seeing it for generations to come.

INKANYAMBA

(INK-an-yam-bah)

FIRST REPORTED	800 BCE
LOCATION	SOUTH AFRICA
CRYPTID TYPE	RIVER MONSTER
REALITY RATING	★ ★

FACTOID: In 1995, Bob Teeny shared a picture of a long-necked dinosaur he claimed was Inkanyamba. It was a hoax—a lie meant to deceive people.

EYEWITNESS ACCOUNT: Inkanyamba (also spelled Nkanyamba) translates to "tornado serpent." Legends of the Zulu peoples of the region say the Inkanyamba lives at the base of Howick Falls in South Africa. Eyewitnesses say it resembles a giant yellow snake with a horse-like head and blazing red eyes. It is said to fly into

Skull

Adult

the clouds when searching for a mate. And when it is angry, rainstorms are sure to follow.

South African cave paintings show San peoples battling with monumental serpents hundreds of feet long. In some of the ancient paintings, the Inkanyamba is spitting water from its mouth. The San are ancestors of the Zulu, and the stories have been passed down to new generations.

Johannes Hlongwane, a caretaker, claims he saw the Inkanyamba twice—in 1971 and 1981. He said its head was raised from the mist of the falls more than thirty-two feet and a sailed crest ran all the way up its neck.

People who live near the falls warn their children not to swim in the waters below. When their young vanish, they often blame it on the Inkanyamba. Only holy men called "sangoma" are allowed to approach the waters to pray for the lost souls.

But some say the Inkanyamba is not evil. They say it tries to convince people not to give up on life and guides lost souls to a better place in the afterlife.

Skeptics say the water is too shallow to hide a massive serpent. They say eels and otters are the only predators to swim the waters of Howick Falls. But believers keep warning their kids. And a viewing platform at the top of the falls gives the curious a great place to do their own research.

Baby

ISSIE

(IZ-ee)

FIRST REPORTED	300 BCE
LOCATION	JAPAN
CRYPTID TYPE	LAKE MONSTER
REALITY RATING	★ ★

FACTOID: When Japanese people fell in love with Issie, Lake Kussharo in Northern Japan introduced Kusshi, a sea monster of its own.

EYEWITNESS ACCOUNT: Japanese legends tell a tale of Issie, a beautiful white horse that lived near Lake Ikeda on Kyushu Island with her foal hundreds of years ago. When a samurai kidnapped her foal, Issie searched everywhere with no success. Overcome by sadness, Issie threw herself into the lake. Transformed into a horse-like sea monster, Issie is allegedly still searching for her missing foal to this day.

The legend might have been forgotten. But when twenty people reported an Issie sighting in 1978, interest was renewed. As they attended an event at Lake Ikeda, all twenty said they saw a long black serpent gliding across the water. That same

Adult

year, another witness named Toshiaki Matsuhara snapped a photo of a strange shape in the lake.

The local tourism office had seen Scotland's Loch Ness Monster draw many visitors, so it decided to use Issie in the same way. Issie stuffed animals are everywhere. School children perform plays in Issie's honor. And statues have been erected to honor the horse-turned-sea-serpent.

Some say the sightings are actually Lake Ikeda's famous eels that can grow to be six and a half feet long. Others cling to the legend of Issie with passion and respect.

If you visit Japan's Lake Ikeda and miss the rare appearance of the sea monster, take heart. The lake is in a volcano's caldera, so that's a great reason to visit the lake, even if you don't see Issie.

Skull

Baby

KELPIE/WATER HORSE

(KEL-pee)

FIRST REPORTED	NINTH CENTURY
LOCATION	SCOTLAND
CRYPTID TYPE	RIVER MONSTER
REALITY RATING	★

FACTOID: The Hollywood version in *The Water Horse* was a friendly, harmless lake monster. The early stories were very different.

Skull

EYEWITNESS ACCOUNT: The original water horses were also called Kelpies, a creation of ninth-century Scottish mythology. They have never been real, but the stories are rich in detail, drama, and danger, even today.

Kelpies are solitary water spirits. They can live in water and on dry land. They often appear to be strong black horses that graze near rivers and streams, but they can take human form—human with hooves instead of feet.

In the region of Aberdeenshire, the Kelpie has a mane of wriggling serpents. And the Kelpie of the River Spey is white with a hypnotic singing voice. Some Kelpies are kind and helpful, but most are frightening. More treacherous Kelpies can summon floodwaters to find unsuspecting children. When their tails hit the water, the sound of thunder is heard.

If a child climbs on the back of a beautiful Kelpie, they are glued to the horse forever. If a Kelpie catches more children than its back can carry, it can expand in size by magic.

Some Kelpies carry their captured children into the water. But flashing a Bible is a great defense against a Kelpie. Wave the holy book, and the Kelpie will retreat and look for other children.

A Kelpie's only weakness is its bridle. If any human can take hold of a Kelpie's bridle, the magical spirit becomes powerless. Capturing a Kelpie means claiming a horse stronger than ten typical horses combined.

Some Scottish clans pass down Kelpie bridles from generation to generation.

Kelpies would be a remarkable sight to see. But we're probably lucky this cryptid is not real.

Baby

Adult

KRAKEN

(CRA-kin)

FIRST REPORTED	1180
LOCATION	DENMARK/NORWAY/SWEDEN
CRYPTID TYPE	SEA MONSTER
REALITY RATING	★ ★ ★ ★

FACTOID: The giant squid—a beast that probably inspired the stories of the Kraken—has the second largest eye on planet Earth. It scours the seas for sperm whales, the only predator capable of eating a giant squid.

EYEWITNESS ACCOUNTS: In the earliest stories, the Kraken was described as a giant cephalopod capable of dragging giant ships and their crews to the depths of the ocean floors. The tales were not limited to Denmark, Norway, and Sweden. Similar stories were told in ancient Greece, Japan, and New Zealand.

Because squid and octopus bodies are so different from landlocked animal species, it was easy to imagine them as monsters called Kraken.

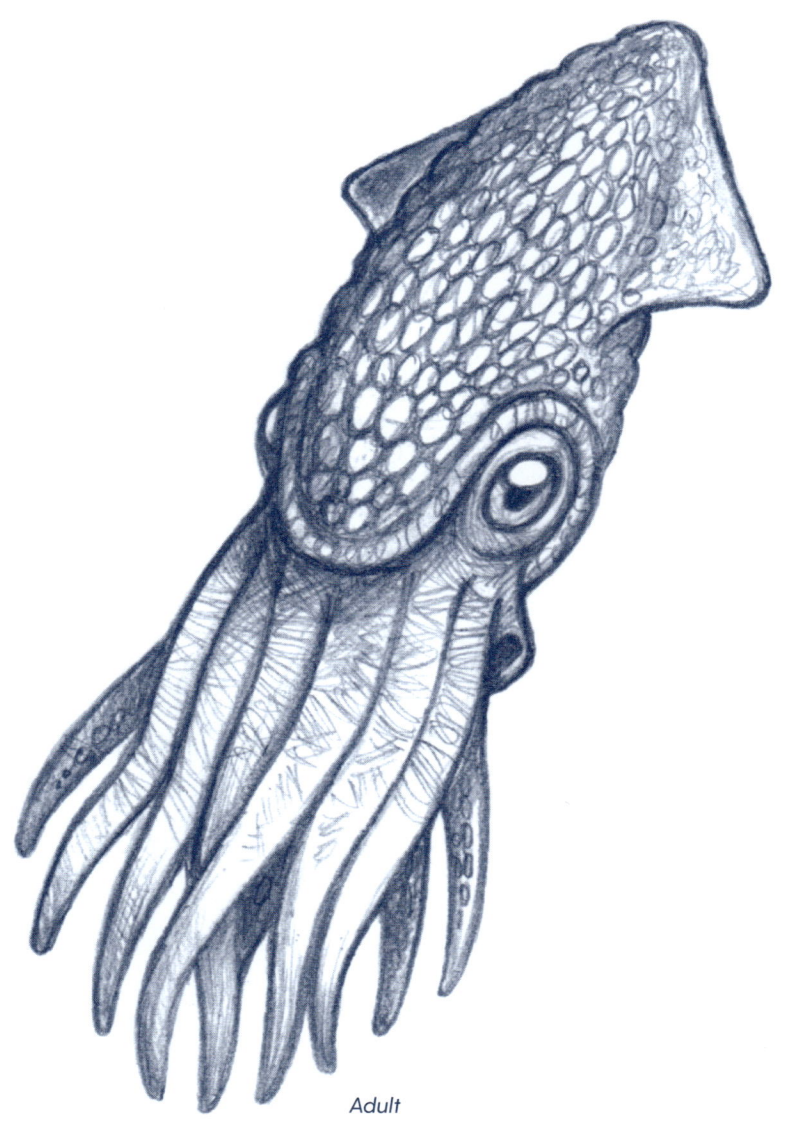

Adult

Baby

When scientists confirmed the giant squid did exist in 1857, the mysteries began to come together. The giant squid lives in the depths of temperate waters all over the world. Females are up to 43 feet long. Males are a bit smaller, at 33 feet.

Marine biologist Dr. Edie Widder created a giant squid lure. It mimicked the bioluminescent light a jellyfish emits when it's being attacked by a larger predator in the deepest waters of the ocean. She believed that light attracted the squid, which wanted to eat the creature preying on the jellyfish.

When she tested the lure from inside a glass submersible vehicle in 2012, she was proven correct. Deep in Japanese waters, a giant squid appeared, hoping for a bite to eat. The video she captured was grainy but astonishing.

Dr. Widder believes thousands of the enormous predators swim the depths of the ocean. Their sharp, beak-like mouths are lined with razor-sharp teeth to devour any living sea creature they encounter—even their own kind. But ships are not on their meal plan. And while smaller squid species hunt in packs, giant squid are loners that avoid human interaction.

That solitary lifestyle has made it hard to study the giant squid. But marine biologists like Dr. Widder are determined to learn more in the future. So if Kraken are actually giant squid, they are real life sea monsters just waiting to share more of their secrets.

Beak

LAGARFLJÓT SERPENT/ LAGARFLJÓT WYRM/WORM

(LA-ger-float)

FIRST REPORTED	**1345 OR 1589**
LOCATION	**ICELAND**
CRYPTID TYPE	**LAKE MONSTER**
REALITY RATING	★ ★ ★ ★

FACTOID: In 1589, the Lagarfljót Serpent allegedly raised its hump so high, ships could pass under it.

EYEWITNESS ACCOUNT: An Icelandic folktale told of a slug living with a rich man's gold. It grew into a powerful monster and moved to Lagarfljót to swim without limitations.

Now known as the Lagarfljót Serpent or Lagarfljót Worm, it has appeared on ancient maps, in famous poetry, and in historic records. In the earliest eyewitness stories in both 1345 and 1589, the serpent raised its hump so high, it caused

Adult

earthquakes when it slapped back down on the surface of the lake.

As the centuries passed, the sightings continued. Said to be between ten and forty feet long, it's been seen in and out of the water. One minute it can be seen coiled around a tree, and the next, slithering back into the water.

In 1963, the head of the Icelandic Forest Service saw the monster. Twenty years later, workers laying telephone cables saw it near the eastern shore of the lake. When they examined the cable where the serpent had been, it had been destroyed—crushed by the serpent's powerful jaws or sharp scales.

In 1998, a teacher and her students spotted the serpent on a field trip to the lake.

Thanks to increased interest in the monster, boat tours began in 1999 to search for the Lagarfljót Serpent. Regional tourism organizations hope to use the interest to attract even more visitors from all over the world. Stacks of witness statements have been compelling throughout the years, but when video evidence surfaced in 2012, it reinforced the older statements.

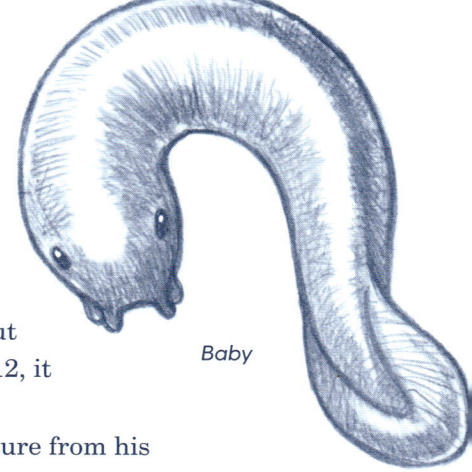

Baby

Hjörtur E. Kjerúlf saw the creature from his kitchen window and thought it might make an interesting video. When he shared it with his local television news station, they posted it on YouTube. Millions of viewers made Kjerúlf's video go viral.

Not everyone believes the video is legit, but Kjerúlf insists, "This is no joke." And he never actually said it was the Lagarfljót Serpent. He simply shared a surprising video and left it to viewers to draw their own conclusions.

Is a giant serpent swimming through the lake's often icy waters? Without more evidence, we'll have to leave that decision to you too.

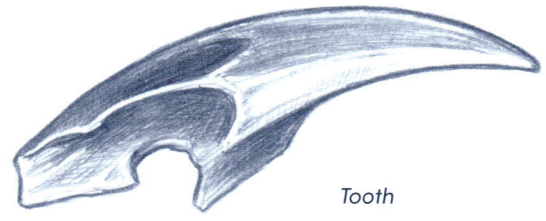

Tooth

Centre for Fortean Zoology

In 1992, British researchers formed the Centre for Fortean Zoology (CFZ). It is a scientific organization in the United Kingdom dedicated to cryptozoology—the study of unknown animals. Their mission has four points of focus:

- To investigate and research mysterious animals around the world in a scientific process

- To investigate reports of strange animals thought to be extinct or beyond their natural habitats

- To publish all findings and make that information available to all readers—full transparency

- To promote environmental justice, conservation, sustainability, and animal welfare

The CFZ produces a weekly video show that covers cryptozoological news in an entertaining way. It also publishes books and magazines.

Richard Freeman is the zoological director of the CFZ and the host of the CFZ's television series. His obsession with all things weird began with a passion for science fiction books and the BBC program *Doctor Who*. He went on to study zoology at Leeds University, then he became a zookeeper at the Twycross Zoo in Leicestershire, England. Today, he works full time at the CFZ, writes and speaks about cryptozoology, and continues to search for the truth about animals, known and yet to be discovered.

You can read more at cfz.org.uk.

LAKE CHELAN DRAGON

(SHA-laan)

FIRST REPORTED	1600s
LOCATION	WASHINGTON, UNITED STATES
CRYPTID TYPE	LAKE MONSTER
REALITY RATING	★ ★

FACTOID: The Indigenous peoples of Lake Chelan—the Chelan and Squaxin—were probably the first to witness this lake monster.

EYEWITNESS ACCOUNT: Ancient legends warned that traveling across the waters of Lake Chelan could be dangerous. There was an evil spirit just beneath the surface. When it took the form of a giant serpent, it could wipe out homes and livestock.

Those same historic tales spoke of a kinder spirit that tried

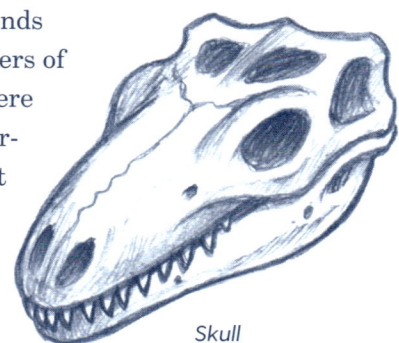

Skull

Baby

to control the evil serpent by blocking river exits, hoping the beast would drown. This attempt only made the Lake Chelan Dragon angry enough to wipe out the whole village.

One girl survived, according to the tales. She was gathering berries in the woods when the monster slaughtered her people. Trapped, she painted her life story on the rocks of Stehekin before she died. Those petroglyphs may still exist in Chelan.

Another story says the Lake Chelan Dragon came to Washington State from Scotland in 1812. When the crew loaded a locked chest onto the ship, they assumed it was filled with gold and other treasures. They were wrong.

When a storm hit the ship, it was magically transported to Lake Chelan, where the heavy chest was washed overboard. The ship's captain dove in to save the riches, but the

weight of the chest was too much for him. When two women leapt in after the captain, they were luckier.

According to witnesses, the women were transformed into Mermaids the minute they hit the water. One rescued the captain. The other saved the chest—and the dragon inside it—the Lake Chelan Dragon.

The dragon may have resurfaced in 1892, when a pair of teenagers swimming in the lake heard a friend cry out in pain. Razor-sharp teeth had a death grip on their friend's leg beneath the water, and they feared he would drown. The two friends fought to save the injured boy, dragging him back to the shore along with the monster.

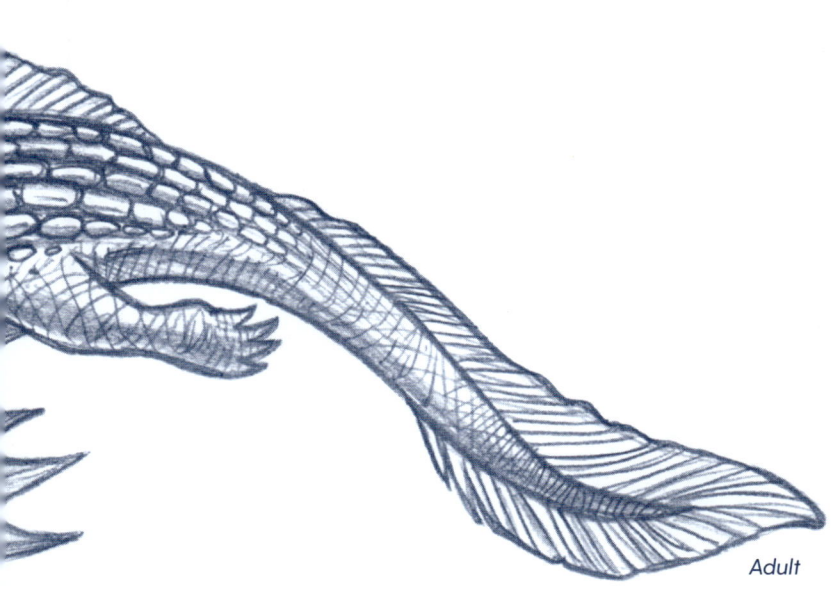

Adult

The two boys said the creature had the legs and body of an alligator, the head and eyes of a snake, scales across its entire body, and bat wings. While they described the creature's skin as "soft as velvet," the creature seemed impossible to kill. Sticks, rocks, and knives were useless. Fire finally frightened the dragon enough that it flew away with the injured swimmer still in its jaws. He was never seen again.

Are the stories true? Is the Lake Chelan Dragon still threatening people who dare dip their toes in the water? It's hard to say. Lake Chelan is the deepest lake in Washington, its bottom has never been explored, and the dragon has been a popular legend for centuries.

LAKE VAN MONSTER

FIRST REPORTED	1995
LOCATION	TURKEY
CRYPTID TYPE	LAKE MONSTER
REALITY RATING	★ ★ ★

FACTOID: Lake Van was once known as the upper sea, according to Tahsin Ceylan, a university professor who searched for the Lake Van Monster in 2017.

EYEWITNESS ACCOUNT: Unal Kozak is also on a mission. He hopes to prove the Lake Van Monster is real. His curiosity was piqued when he worked at Van Yüzüncü Yıl University and heard the first eyewitness stories in 1995.

Lake Van is a saltwater lake and is the largest lake in Turkey, but it's relatively shallow with its deepest section only 1,467 feet deep. Could it hide a mysterious monster? Kozak mapped

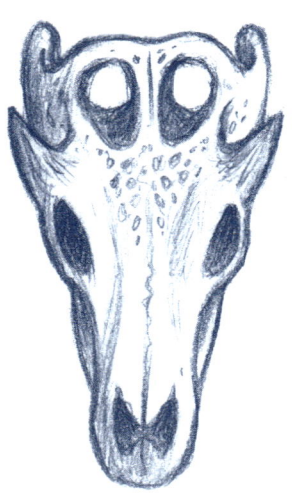

Skull

Adult

Lake Van's monster locations based on eyewitness interviews then began to visit each spot himself. Three times, Kozak's hard work paid off and he saw the creature too. And three times, he filmed it gliding across the water.

According to those film clips, the dark-colored monster is nearly fifty feet long with triangular spikes on its back. Small horns protrude from its head, which is sparsely covered with what looks like hair.

Kozak continued his study, interviewing thousands of witnesses and eventually wrote a book on the Lake Van Monster. Inspired by Kozak's success, film crews from France, Japan, and other international locations came to Lake Van in hopes of filming the monster, with very little success.

Van Yüzüncü Yıl University mounted an underwater search in 2017 to confirm Kozak's evidence. Ceylan led the team, hoping to find a creature similar to the Loch Ness Monster. "They told us that we won't find anything," he said. But they were wrong.

It's true that Ceylan found no proof the monster was real. But he did find an ancient castle from the sixth century. And he's still searching for a lake monster to go with it.

Baby

LARIOSAURO

(LAR-ee-oh-SAHR-oh)

FIRST REPORTED	**1946**
LOCATION	**ITALY**
CRYPTID TYPE	**LAKE MONSTER**
REALITY RATING	★ ★

FACTOID: In 1830, naturalist Giuseppe Balsamo Crivelli discovered a fossilized *Lariosaurus* near Lake Como. It was a prehistoric reptile two feet long.

Skull

EYEWITNESS ACCOUNT: In 1946, the Italian newspaper *Corriere Comasco* ran a startling story. A giant water monster had been spotted at Pian di Spagna, a nature reserve on the northern side of Lake Como. It was the first of many creature sightings.

That first account described an animal between thirty and forty feet long with red scales and a reptilian head. The two hunters that spotted it had no time to raise their rifles. It disappeared almost as quickly as it appeared.

Adult

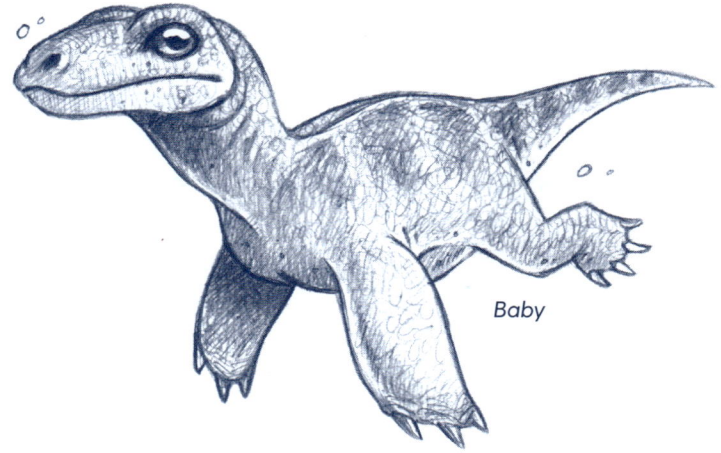

Baby

In 1954, a father and son saw a much smaller mystery creature on the western bank of the lake. They said it was about three feet long with a rounded head and webbed feet.

By September of 1957, biologist Renzo Pagani had turned his sights on the Lake Como mystery. He decided to create and launch a bathysphere—a primitive round submarine—in search of the lake monster. He found an underwater reptile about seven feet long with a crocodilian head.

Almost fifty years later, a Lake Como fisherman spotted a forty-foot-long eel near the eastern shore in 2003.

Because each sighting is different, the witnesses probably did not see the same animal in Lake Como. And no sketches or photographs of the creatures exist. But the sighting in 1946 is the only one that would qualify as a possible lake monster.

Some claim the lake monster is a Lariosaurus that escaped extinction and grew very large, but that is unlikely. The search for more evidence continues.

LOCH NESS MONSTER

(LOCK ness)

FIRST REPORTED	**SECOND CENTURY**
LOCATION	**SCOTLAND**
CRYPTID TYPE	**LAKE MONSTER**
REALITY RATING	★ ★ ★ ★

FACTOID: The Picts, an ancient people with no written language, first carved the image of the Loch Ness Monster in stone in second-century Scotland.

EYEWITNESS ACCOUNT: For centuries, the people who lived in the land now known as Scotland believed a large animal lived in Loch Ness—a deep, cold lake carved by an ancient glacier.

The first written account of the Loch Ness mystery came in the year 565 in the biography of Saint Columba, a man who brought Christianity to Scotland. On his way to speak with the Pictish king, he saw a fearsome beast about to attack a man in the water. Columba commanded the creature to retreat and saved the man.

More sightings were reported after a road was built near the shore in 1933. One couple reportedly saw the beast

Adult

"thrashing about" in the lake. A reporter who wrote about their story for Scotland's *Inverness Courier* used the term "monster," and it stuck.

Skull

A second couple saw the Loch Ness Monster crossing the road, complete with visible flippers, and the race was on. Newspapers all over Europe sent reporters to Scotland to cover the mystery. The public demanded more information about the creature they now called Nessie.

The *London Daily Mail* hired a big-game hunter named Marmaduke Wetherell to track down the Loch Ness Monster. He quickly found footprints of a huge four-toed beast he said proved the monster was real. But they were a hoax.

Burned by the lies, experts stopped investigating, but the eyewitness reports continued—more than four thousand from the 1930s to the 1950s alone—all compiled in a book called *More than a Legend* by Dr. Constance Whyte in 1957.

Oxford University, Cambridge University, and the University of Birmingham soon launched serious scientific studies. When the Academy of Applied Science in Boston, Massachusetts, joined them in 1975, physicist Robert Rine proposed using sonar, underwater cameras, and a flashing strobe light as they searched for new clues. The new tech paid off. Cameras captured what appeared to be a marine reptile with flippers.

Rine led another exploration in 1997, and the documentary television series *NOVA* went along to film the results. Rine captured more puzzling images on sonar—of what looked like a marine animal fifteen feet long—the size of a whale or a Nessie.

In 2019, DNA specialist professor Neil Gemmell collected 250 water samples from the lake but found no plesiosaur DNA. What he did find was eel DNA, and a lot of it. Could Nessie be a supersized species of eel?

The evidence is inconclusive, but just because you can't prove a fact doesn't mean it's not true. Perhaps someday, science will crack the Nessie code once and for all.

Baby

Loch Ness Monster x2

In March of 2023, monster hunter Eoin O'Faodhagain was monitoring his webcam focused on the Scottish Loch Ness. Sightings had been scarce, and he was hoping for evidence that Nessie was still there.

His excitement mounted when he spotted a huge black shape gliding across the lake. At first, he thought it had two humps—one big and one smaller. But when he saw the bigger hump moving away from the little one, he was astonished. He had to ask the question: is it possible that Nessie is not alone?

The first creature was dark in color and about thirty feet long, according to O'Faodhagain. And there were no other breaks in the surface of the lake besides the two he witnessed. He was convinced—there were two mysterious animals coexisting peacefully in the loch.

O'Faodhagain admits that there are many large eels living in Loch Ness. He even admits a giant eel could explain some Loch Ness Monster sightings. But if there are two different unknown marine animals in the lake, it could account for why some people describe Nessie in different ways.

"This is only my opinion," O'Faodhagain told the British newspaper *The Mirror*. But it's an opinion that might explain at least part of the Loch Ness mystery.

Adult

LUSCA

(LOU-ska)

FIRST REPORTED	**1896**
LOCATION	**CARIBBEAN ISLANDS**
CRYPTID TYPE	**SEA MONSTER**
REALITY RATING	★ ★

FACTOID: Lusca is an illustrated character in *Pathfinder*, a role-playing game. It has three shark heads in the front of its body and a tangle of octopus tentacles in the back.

EYEWITNESS ACCOUNT: Blue holes are common in the ocean waters surrounding the Caribbean Islands. Because they are deeper than the rest of the waters, they have a darker blue color—that's why they are called blue holes. These deep, mysterious blue holes can be dangerous for swimmers.

Currents near the holes can be unpredictable and can suddenly pull swimmers into the depths. Some blue holes are also starved of oxygen, which can be disorienting for scuba divers. And the warm, shallow waters grow cold in the deep blue depths, making it harder to swim.

Those dangers probably inspired the terrifying stories of Lusca sea monsters. Island parents might have hoped the monster stories, like the fictional boogeyman, would keep their kids away from the holes. Or the stories may reveal a true cryptid.

Luscas are described as bloodthirsty marine predators—half great white shark and half giant octopus. The shark end has a never-ending supply of razor-sharp teeth. The octopus end is a swarm of gripping tentacles, made for capturing prey.

Luscas are between seventy-five and 250 feet long and can change color like a cuttlefish (an octopus relative). As they hide in silence, camouflaged to their surroundings, they wait for a swimmer to approach then strike—alone or in groups.

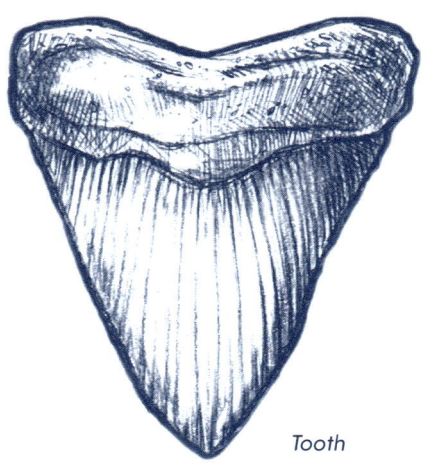

Tooth

Baby

When they are about to attack, a storm of bubbles rises out of the blue hole. By the time you see the bubbles, it's too late. Luscas are very fast. And if you make it to shore, they can follow you onto the beach to make the kill.

In 1896, two teenage boys discovered the lifeless body of a Lusca washed up on Anastasia Beach in Florida—complete with tentacles. Scientists disputed the Lusca theory and said it was a simple octopus that washed ashore.

Jeremy Wade tried to catch a Lusca in an episode of his *River Monsters* television series. He was not successful, but he admitted something unknown could have caused a number of blue hole drownings in the Caribbean.

MANIPOGO

(MAN-ee-POE-go)

FIRST REPORTED	1800s
LOCATION	MANITOBA, CANADA
CRYPTID TYPE	LAKE MONSTER
REALITY RATING	★ ★ ★

FACTOID: The cryptozoologists at the Centre for Fortean Zoology call Manitoba, Canada, a wonderland for cryptids. In addition to Manipogo, the easternmost prairie province claims Waheela (a huge wolf), Sasquatch, and other mysterious creatures.

EYEWITNESS ACCOUNT: The original stories of Manipogo began with the First Nations peoples hundreds of years ago. But witnesses have made similar claims ever since.

Described as being a serpent between twelve and forty-five feet long, Manipogo is brown and black in color and has a head that resembles that of a horse. Some say it makes a thunderous cry when it breaks the surface of the water.

Tom Locke saw Manipogos in August of 1960 when he and sixteen other people spotted three of the serpents

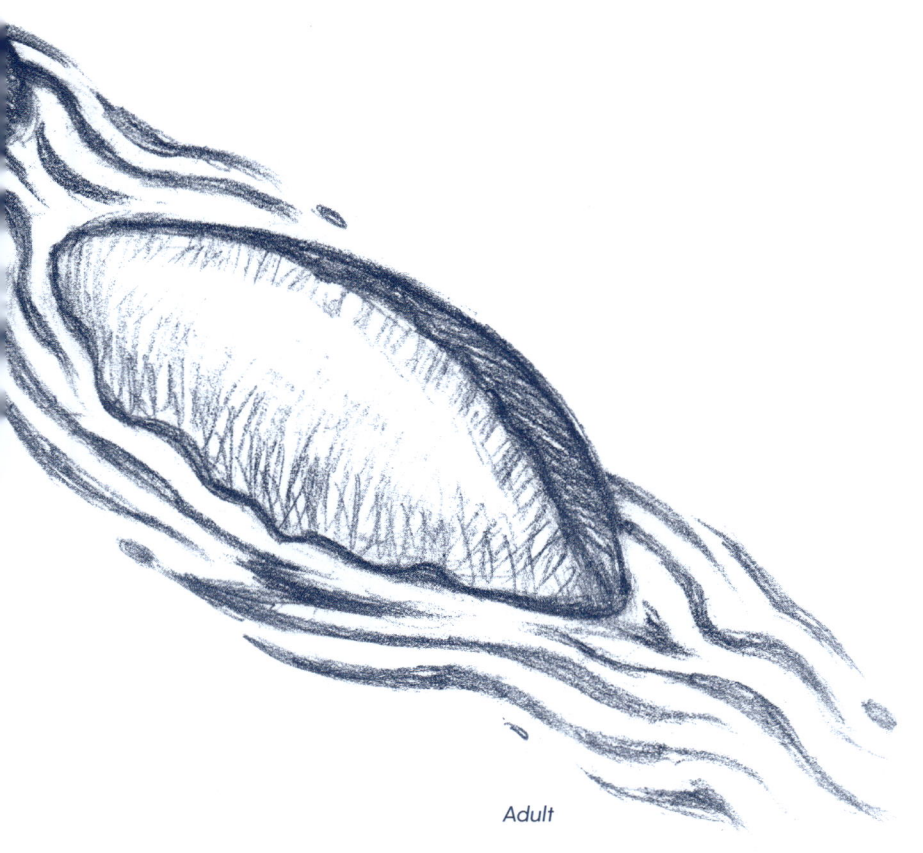

Adult

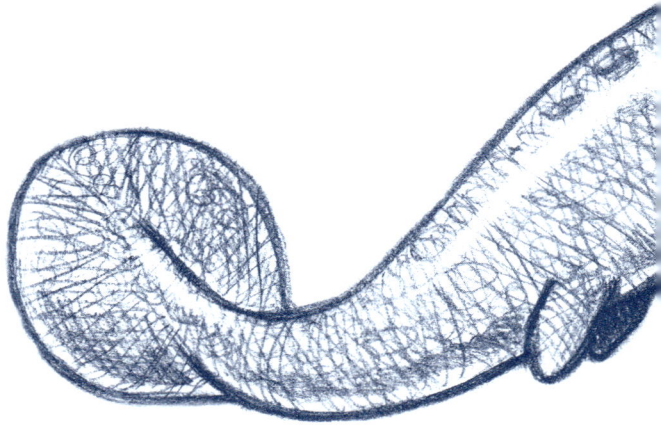

swimming together on the northwest shore of Lake Manitoba. That was the year the lake monster got the name Manipogo. It was a mix of Manitoba—the Canadian province where it allegedly lives—and Ogopogo, another Canadian sea monster in British Columbia.

Two years later, reporter Dick Vincent and his friend John Konefell sputtered through the waters in their small motorboat. They were stunned when they spotted Manipogo's head just above the lake's surface. Vincent and Konefell tried to chase the lake monster, but it was too fast. They did snap a picture—a large, blurry head and neck, about two feet above the water. They said it had at least one hump and was almost forty feet long.

Sightings continued, but in June of 1997, a farmer harvesting hay made an unbelievable claim. He said he not only saw Manipogo, he shot and killed it with his rifle. But

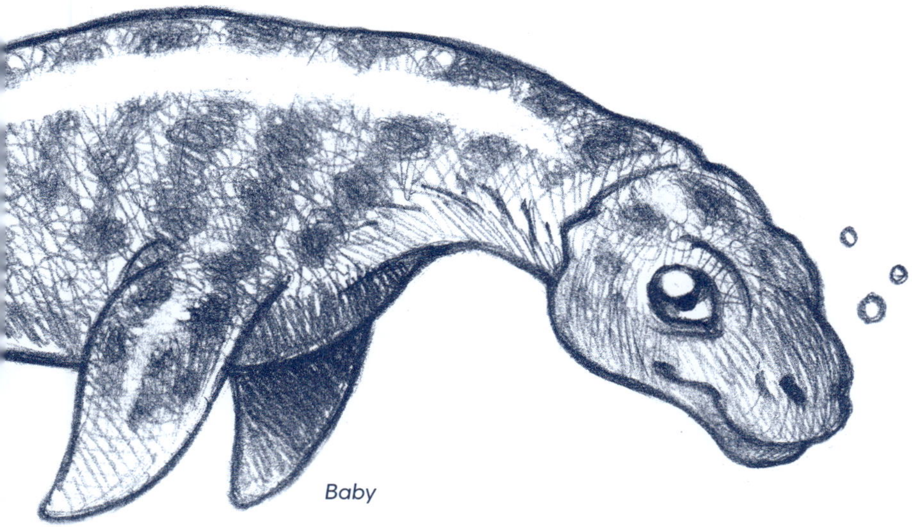

Baby

when he couldn't produce the body, people assumed the tale was untrue.

Keith Haden had a more credible experience in 2004. The commercial fisherman had hauled in his nets hundreds of times before. But one day that year, his nets came out of the lake ripped to shreds. Haden was sure a large predator had done the damage—perhaps a Manipogo.

Author Chris Rutkowski collected dozens of Manipogo stories for his book, _Unnatural History: True Manitoba Mysteries_. He admits stories aren't scientific proof. But he plans to keep looking until DNA proves Manipogo is really out there.

Skull

MEMPHRE

(MEM-free)

FIRST REPORTED	**1926**
LOCATION	**VERMONT, UNITED STATES AND QUEBEC, CANADA**
CRYPTID TYPE	**LAKE MONSTER**
REALITY RATING	★ ★ ★

FACTOID: Memphremagog is the Abenaki word for "great water plain."

EYEWITNESS ACCOUNT: Ancient stories have told of a monster in Lake Memphremagog for centuries. The Indigenous peoples warned newcomers not to swim in the dangerous waters. Official written reports surfaced decades later.

Novelist Norman Bingham fictionalized the stories in his book *The Sea Serpent Legend* in 1926. He described a dark monster with a tusk and a horn swimming across the lake in a "coiling surge."

In 1961, two fishermen spotted a black creature about forty feet long swimming past their boat. In 1994, four people in two boats saw a thirty-foot-long black animal with three visible humps.

By 1997, University of Sherbrooke researcher Sonia Bolduc had collected 215 well-documented sightings of the monster, now called Memphre. And in 2022, a witness captured the monster on video. According to Bolduc, most witnesses say Memphre resembles the Loch Ness Monster.

A second historian, Jacques Boisvert, actively searched for Memphre evidence by scuba diving. He did underwater investigations in Lake Memphremagog seven thousand times before he passed away in 2006. His loved ones said Boisvert

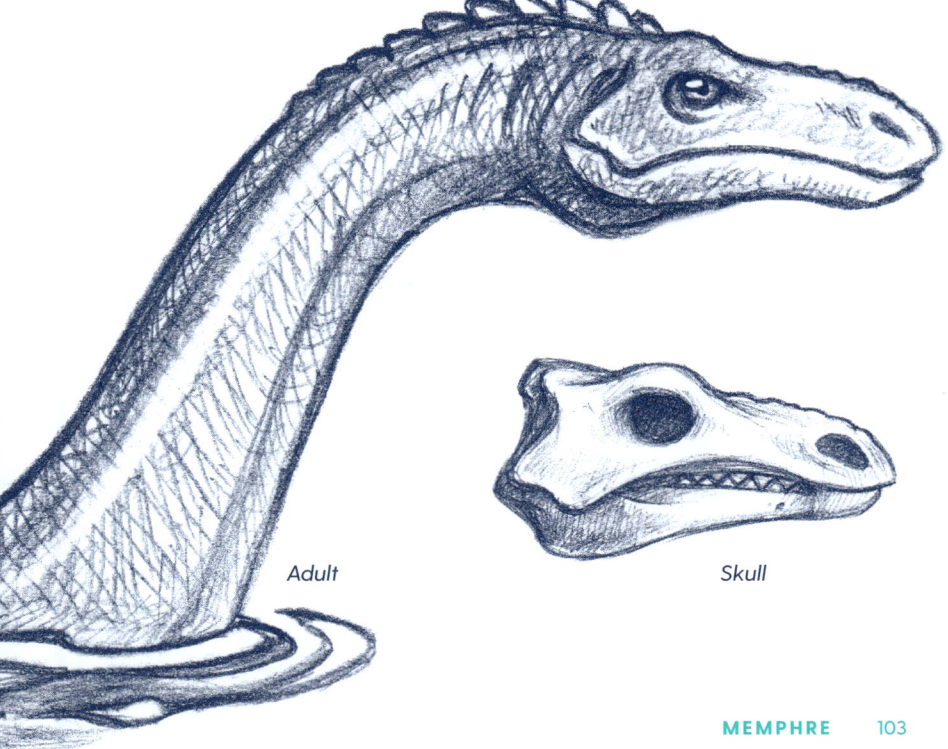

Adult

Skull

Baby

always kept an open mind while playfully calling himself a "dragonologist."

Barbara Malloy, the head of the Memphremagog Historical Society of Newport, insists there are Viking petroglyphs near the lake that feature Memphre—ancient works she's seen with her own eyes.

When she first saw the actual lake monster, Malloy thought it was a person on a motorized watercraft because it was traveling so quickly. But when she spotted the legendary horse-shaped head, the long neck, and the thick body, she knew it was Memphre. According to Malloy, it vanished almost as quickly as it appeared.

She believes the serpent lives at the bottom of the lake. She thinks Memphre is intelligent and evasive, as are so many lake, river, and sea monsters. Those hoping to capture a glimpse of the shy lake monster should start at the official Lake Memphremagog lookout station dedicated to Boisvert.

Royal Canadian Coins

In 2011, the Royal Canadian Mint created a series of mythical creature coins. The Memphre was one cryptid featured, in honor of the monster of Lake Memphremagog, ninety-three miles southeast of Montreal.

Dracontologist Jacques Boisvert helped to create the artful coins that came in a fully illustrated color folder with a pullout map marking the locations of mystery animal sightings in Canada.

Each coin is made of nickel-plated steel. The Memphre coin has a silver-colored background and a colored reproduction of the serpent's head and snout on one side. On the other side is Queen Elizabeth, the monarch who reigned in Great Britain and Canada when the series was created.

The face value of the coin is twenty-five cents—a Canadian quarter. But they cost roughly twenty-five dollars each when they first went on sale.

Other creatures featured on the cryptid coins include Mishipeshu, the "Great Lynx" shape-shifter of Lake Superior, and Sasquatch, the mysterious dark-colored biped that allegedly lives in the rustic wilderness worldwide.

Mermaid

MERMAIDS/MERMEN/MERFOLK

FIRST REPORTED	1,000 BCE
LOCATION	WORLDWIDE
CRYPTID TYPE	SEA MONSTER
REALITY RATING	★

FACTOID: P. T. Barnum's Feejee Mermaid was not a Mermaid. It was a hoax. It was half a monkey sewn to half a fish and mummified.

EYEWITNESS ACCOUNT: In 2012, Animal Planet and Discovery Channel featured *Mermaids: The Body Found*. They later aired a sequel, *Mermaids: The New Evidence*, one year later. Both channels normally share documentaries—true stories. But these were fictional. They introduced a scientific team said to be working for the National Oceanic and Atmospheric Administration (NOAA). That team said Mermaids were real.

A so-called marine biologist named Dr. Brian McCormick said he discovered the partially digested body of a Mermaid inside a whale carcass in South Africa. But the evidence was allegedly stolen before he could finish his analysis.

It sounds amazing, but here's the problem: McCormick did *not* work for NOAA. He did not exist. He was a fictional character played by an actor named Sean Cameron Michael.

The programs shared a six-thousand-year-old sandstone cave painting of Mermaids. But they were actually from the Egyptian Cave of Swimmers rock art featuring human beings. The human swimmers were turned into Mermaids as a deception.

Animal Planet and Discovery Channel added a note, calling the Mermaid programs "speculation" created for entertainment purposes only. But it was at the very end of the show, and most people didn't see it.

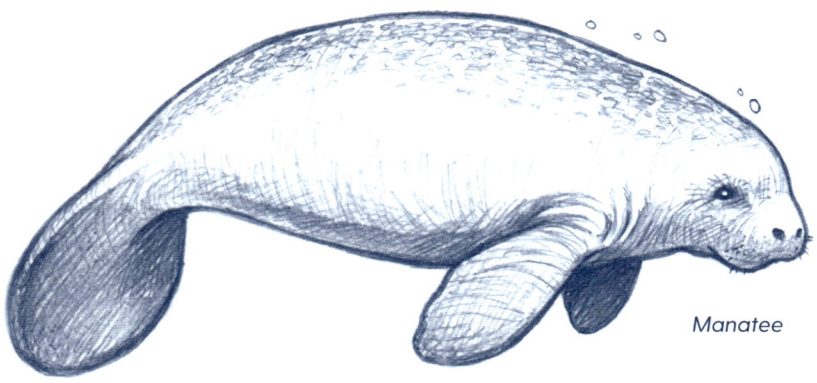

Manatee

Television wasn't the first medium to fake Mermaids. Barnum—the real-life person profiled in the Hollywood blockbuster The Greatest Showman—told the world he had a Mermaid and sold tickets to see it. Even he called it a "humbug," his word for a fake.

Merfolk stories have been part of folklore around the world. Consider Africa's Mother of the Waters, the Sirens of ancient Greece, the Rusalki of Eastern Europe, and the Merrows of Ireland, to name only a few.

The fact that Mermaid stories exist in virtually every human culture on earth is a great reason to explore the realm of Merfolk. But critical thinking can guide you when it comes to the facts and the falsehoods.

Merman

Ningyo, Japan's Amazing Merman

For centuries, citizens of Japan told tales of the Ningyo—a sea creature of Asian origins. But it was not the same creature of European myth. Ningyo was half fish and part human but with the mouth of a monkey. Legends said feasting on a Ningyo could make a person live forever.

Prince Shōtoku, Japan's leader from 574–622 CE, claimed to have seen a Ningyo in Lake Biwa in Kyoto. Shōtoku believed the Ningyo was once a fisherman who had dropped his nets in forbidden waters. Punishment for his crime was being transformed into the Ningyo.

The cursed merman begged the Prince to forgive him and asked for an odd promise. When the Ningyo died, he hoped the Prince would mummify his body and put it on display—a warning against breaking the rules in generations to come.

Today, one so-called Ningyo mummy is cared for by holy men at the Tenshou-Kyousha Shrine in Fujinomiya. Does the shrine harbor the original cursed fisherman or a replica? It's hard to know for sure. But similar mummified "Merfolk" came to the United States from Japan, including those revealed by P. T. Barnum.

Another Ningyo is on exhibit at the Clark County Historical Society in Ohio. Historians passed it to scientists at Northern Kentucky University for study. They confirmed the Ohio Ningyo was half monkey and half fish, with the mummified claws of a lizard attached to its artificial arms. What lizard? The researchers suspect they came from a Komodo dragon.

MISHIPESHU

(MEE-shee-peh-shoe)

FIRST REPORTED	**1600s**
LOCATION	**MICHIGAN, UNITED STATES AND ONTARIO, CANADA**
CRYPTID TYPE	**LAKE MONSTER**
REALITY RATING	★ ★

FACTOID: Some say the famous shipwreck of the SS *Edmund Fitzgerald* was caused by the Mishipeshu.

EYEWITNESS ACCOUNT:

The Anishinabe, the Ottawa, the Ojibwa, and the Potawatomi may have been the first human beings to witness Mishipeshu. But the first written report came from the European priest, Father Paul Le Jeune, in the 1600s.

Skull

Adult

As the priest watched the Indigenous people fishing on Lake Superior, he saw the waters turn turbulent, just as the men pulled something strange out of the water. He said it was almost five feet long with a four-legged reptilian body and the head of a turtle.

Alarmed, the fishermen quickly tossed the creature back into the lake. Did the strange creature cause the rough waters? Did the fishermen save their own lives by setting it free? It's hard to know, but more stories of Mishipeshu soon followed—stories that changed the way we imagine it.

The priest may have seen an odd lizard/turtle hybrid, but most Indigenous traditions Ojibwa, Algonquin, Ottawa, Menominee, Shawnee, and Cree—consider Mishipeshu a water panther or a water lynx, with reptilian qualities. It has the head and the paws of a mountain lion and the horns of a buffalo. Its feline body is covered in reptilian scales or feathers. Spikes protrude from its spine and a long tail completes the monster's body.

Mishipeshu swims the lake and guards a cache of copper on Michipicoten Island, a small island about ten miles from the mainland in a place called Thunder Bay. It hides from people in a network of underwater tunnels, roaring and hissing ferocious warnings.

Storms are said to rage when the monster is angry. In 1897, a man fell from his boat into the claws of the aquatic cat. When he broke free, three friends rushed him to shore, but his life was forever changed. To him, Mishipeshu was very real.

Baby

When hiker Randy Braun came to Michigan in 1977, he planned to explore the Porcupine Mountains Wilderness State Park. But when the trails led him to the lake, something unexplainable was waiting for him. Mishipeshu rose from the waves. Fearing for his life, Braun hid behind a boulder and pulled out his camera. He captured one photograph before the monster rushed him. Then he escaped. The picture didn't look like a water panther. It looked like a giant snake with the head of a horse, large dark eyes, and the whiskers of a catfish.

But some say Mishipeshu is a shapeshifter. If that's true, it could take almost any physical form and remain a mystery forever.

MONSTER OF BLACK WATER

FIRST REPORTED	**1933**
LOCATION	**ITALY**
CRYPTID TYPE	**LAKE MONSTER**
REALITY RATING	★ ★

FACTOID: Mysterious creatures—from giant octopus to Merfolk—have been featured in Italian works of art for centuries. The Monster of Black Water joined that grand tradition in 2011.

Skull

EYEWITNESS ACCOUNT: Some say the Monster of Black Water, a sleek, black serpent, has been swimming under the Punta della Dogana on the Grand Canal in Venice for hundreds of years. It hides in the deepest places and tries to avoid the city's crowded piers and walkways most of the time.

One thing is sure to bring the serpent to the surface—complete darkness. When the moon disappears and the night is as dark as possible, the monster's noble head breaks the water's surface. In those black hours, it swims in near silence.

The most recent eyewitnesses came forward in 1933. Two fishermen trying to harvest cuttlefish (related to octopus) were surprised to see the Monster of Black Water leap out of the water to capture and devour a seagull in a single gulp. Few stories have surfaced since.

But a team of Italian artists paid tribute to the monster in 2011. Architect Simona Favrin designed a twenty-foot skeleton made of black steel. Then she added an array of Murano glass scales, handcrafted by artisan Nicola Moretti and Gianluca Orazioi. Well-placed LED lights add to the haunting appeal, including a pulsing heart and the rise and fall of the serpent's breath. The tribute to the ancient monster was installed in a pool of dark water for all to admire.

When the exhibit first opened in Venice, Italy, in 2011, nearly two thousand visitors admired the Monster of Black Water. It toured the art capitals of Europe for years after its debut. Thanks to this artistic creation, the search for the Monster of Black Water may soon continue.

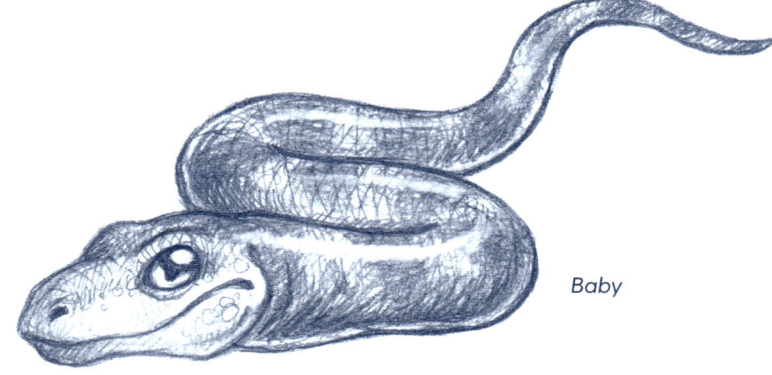

Baby

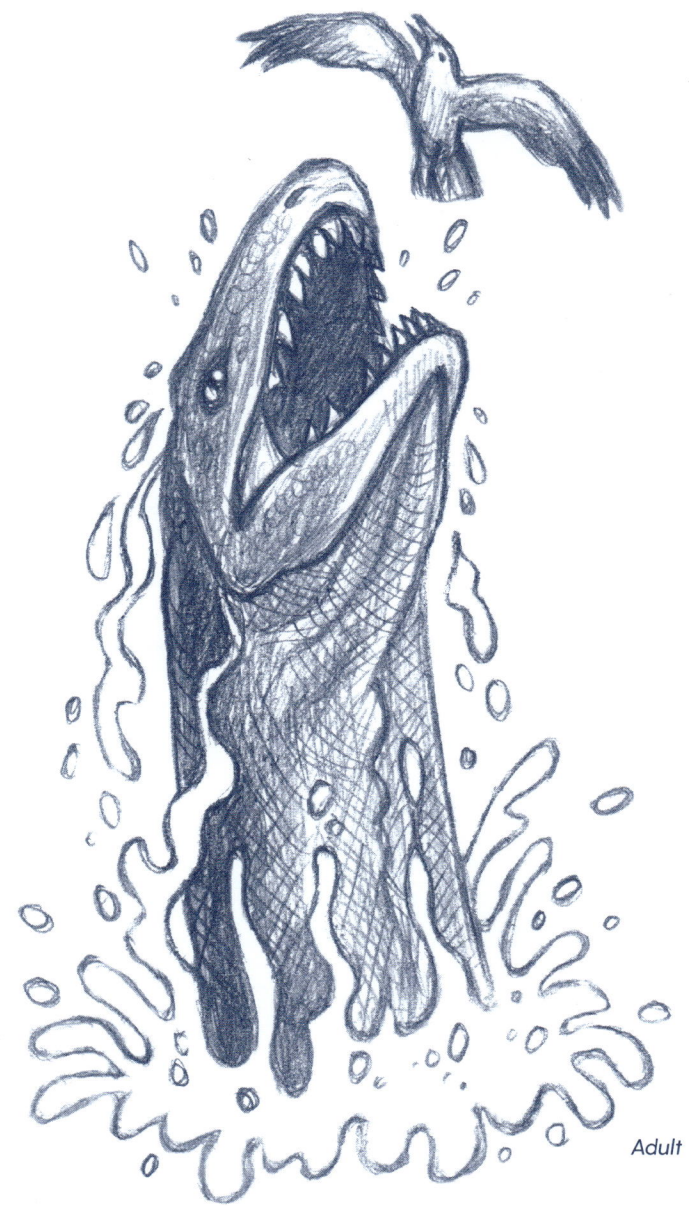

Adult

MONSTER OF OTTAWA RIVER/MAPLE

FIRST REPORTED	**1880**
LOCATION	**OTTAWA, CANADA**
CRYPTID TYPE	**RIVER MONSTER**
REALITY RATING	★ ★ ★

FACTOID: Scientists say most sea, lake, and river monsters are misidentified fish.

EYEWITNESS ACCOUNTS: Some water monsters may be sturgeon, lungfish, oarfish, or eels. But the Ottawa River

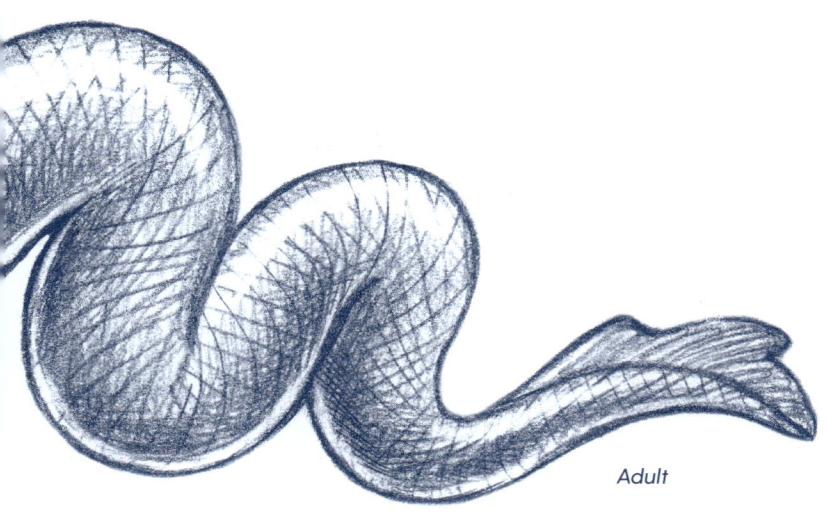

Adult

Monster is a whole other animal—an animal allegedly captured in the 1880s.

At that time, Ottawa, Canada, was booming thanks to the logging industry. Homesteads popped up and steamships were hired to move logs down the Ottawa River to the mills that would turn them into lumber.

When the crew and passengers of the steamship *Levi Young* spotted the monster in 1882, they decided to take action. Newspaper reports said two men jumped into a smaller boat to chase the animal. When they caught up, one man hit the monster in the head with a heavy wooden oar. The creature writhed in pain and turned the water into a frothy sea of bubbles and mud. A second blow from the oar stilled the beast and the men towed it back to the steamship. Once they pulled it on

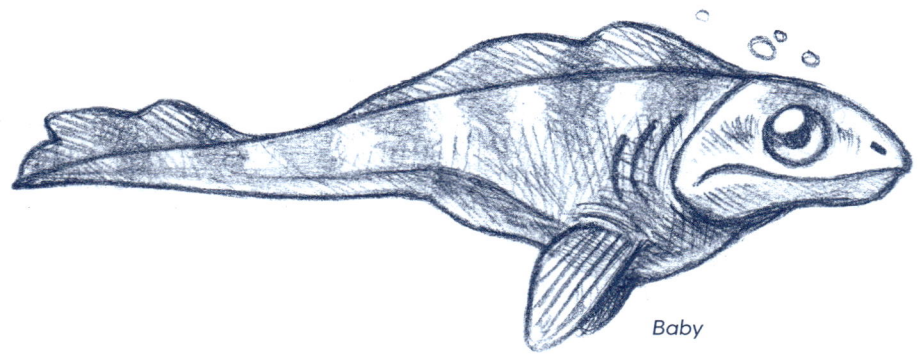
Baby

the ship, they measured it—eleven feet long and more than a foot wide. They also drew pictures, but what happened to the carcass is unknown.

In 2021, Froy Zonn, a college student, was ice skating on the Ottawa River when he heard a rumbling. At first he thought it was his stomach. Then he saw something big, dark, and scaly swimming under the ice. He stepped off the skateway just as the creature broke through.

Zonn wasn't the only witness that night. Half a dozen other people called the Ottawa police to report seeing the same monster. Police refused to investigate, saying monsters weren't in their jurisdiction.

Ottawa natives have nicknamed the creature "Maple," but they are not afraid. Maple has never hurt anyone, so why worry?

Skull

MONSTER OF THE SUSQUEHANNA/ MISIGINEBIG/ SUSQUEHANNA SEAL

(SUS-kwah-han-ah) (MIS-eye-gin-ee-big)

FIRST REPORTED	**1890**
LOCATION	**PENNSYLVANIA, UNITED STATES**
CRYPTID TYPE	**RIVER MONSTER**
REALITY RATING	★ ★ ★

FACTOID: Seals and sea lions, members of the pinniped family, are saltwater marine mammals that sometimes swim in rivers near their oceans. These seals are often misidentified as mystery monsters.

EYEWITNESS ACCOUNTS: When the *Daily Democrat* ran an article about a mystery in the Susquehanna River, the people of Pennsylvania were not all that surprised. Indigenous people such as the Iroquois, Shawnee, Munsee, and Lenape had warned of river monsters for centuries.

In ancient stories about Misiginebig, the serpent was as big around as a tree trunk with a blazing crest on its forehead and horns over its eyes. Both the forehead crest and its full-body scales were said to glow like embers in a fire.

When the logging industry first used the river to move cut trees from one port to another, they claimed to hear "monster calls" along the 444-mile-long waterway. The strange cries caused people to fear the river. But when

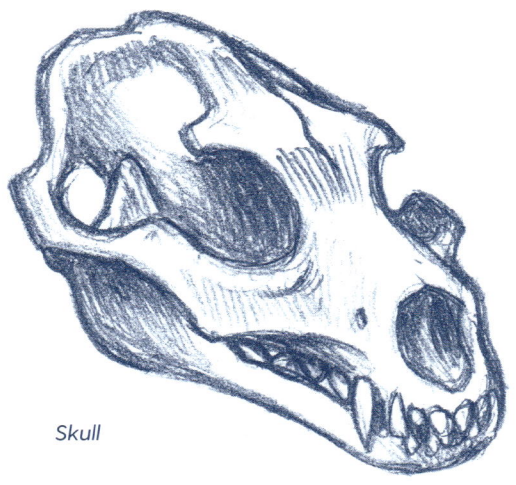

Skull

Adult

people traveling on rafts or canoes suddenly went missing, true terror struck.

Non-Indigenous descriptions differ from witness to witness. Some said it looked like a prehistoric hippopotamus. Some said it looked like a dinosaur. Some agreed with the Indigenous stories, describing a massive snake or serpent.

When paleontologists Ted Daeschler and Neil Schubin discovered a fossilized creature called *Hynerpeton* in 1993, a new theory was born. The fossil looked like a three-foot-long salamander with stubby legs and a tadpole tail. Could the monster be a prehistoric survivor?

A second theory seems more likely. Perhaps the Monster of Susquehanna was actually the Susquehanna Seal—a prehistoric pinniped.

The two theories are hard to reconcile. Seals and salamanders do not look alike. One is a mammal with whiskers and sleek skin-gripping fur. The other is an amphibian with button eyes and slippery, slimy skin. But believers are willing to consider an unlikely hybrid.

If the monster is a seal, a rare seal bite is the greatest danger in the river. If the monster is a seal/salamander hybrid, it's probably not dangerous either. If it's the great serpent Misiginebig, people near the river might want to stay alert, just in case.

Baby

Eyewitnesses say the turtle is as big as the hood of a small car. They say it will chase any child or elderly visitor away if they happen to wander too close when it basks in the sun. No one has captured a picture of the massive turtle, but scuba divers have seen a large rocky-looking creature deep below the water's surface.

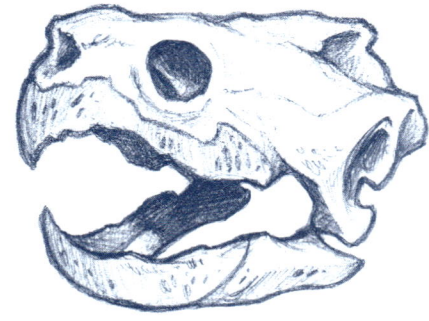

Skull

Experts are doubtful—scientists and cryptozoologists alike. Because the pond was created by city planners and is not a natural body of water, they say it's unlikely to hide a giant turtle. There is no natural history when it comes to a constructed pond. Apart from frogs, animals living within the waters were added by people.

One turtle species common to Mason City is a native snapping turtle about fifteen inches long. But it prefers the movement of streams and rivers to the stillness of lakes and ponds. Even so, some eyewitnesses say this pond monster is an enormous snapping turtle.

The Monster Turtle might be a gimmick—a mystery created to attract tourists to Iowa. If you don't come for a picnic, you might come for a monster turtle. But that's only one theory.

Adult

If you visit Mason City, consider a stop at Lake Okoboji, where a 140-foot-long, bluish-gray cousin to the Loch Ness Monster allegedly lives. It's been said to flip large boats in the water, but it has never eaten the people on board.

Two cryptid mysteries in one trip? That's a vacation that's hard to resist.

Baby

MUGWUMP

(MUG-wump)

FIRST REPORTED	1979
LOCATION	ONTARIO/ QUEBEC, CANADA
CRYPTID TYPE	LAKE MONSTER
REALITY RATING	★ ★ ★

FACTOID: According to dictionaries, "mugwump" means a person who is independent or neutral. But in Ontario and Quebec, "mugwump" means "fearless sturgeon."

EYEWITNESS ACCOUNT: For generations, stories were told of the terrifying monster the Mugwump. It was said to live in the deep waters of Lake Timiskaming in Quebec, and it was enormous.

In the 1920s, John Cobb was moving logs on the lake in a tugboat when he spotted something big. The creature, twenty feet long with a round head and an

Skull

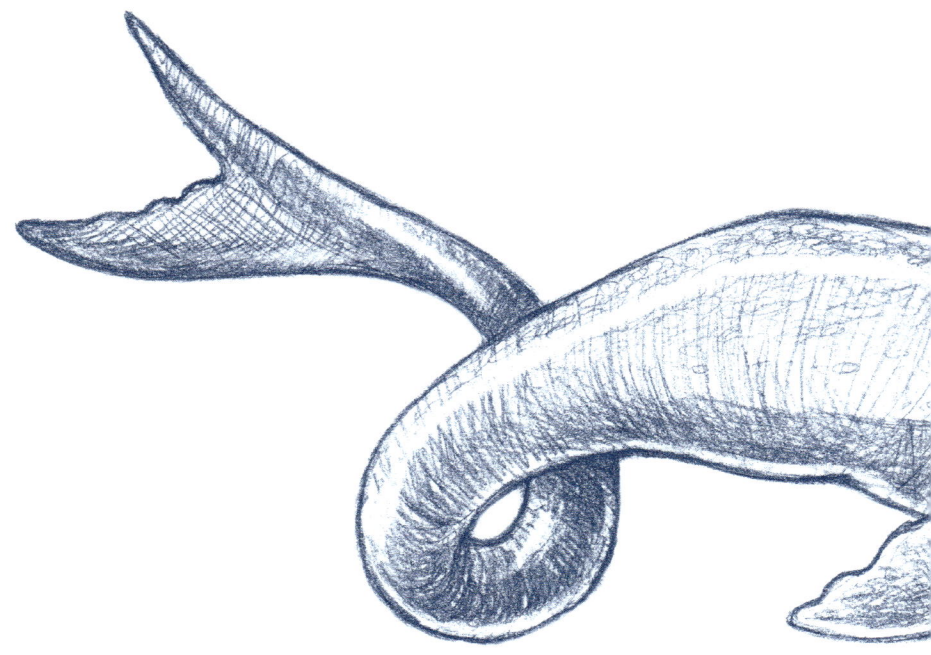

seal-like nose, was right in front of his boat. When he tried to get closer, it vanished.

In 1978, Ernie Chartrand was sitting at a restaurant table watching the lake when he noticed something strange. A large creature was headed right toward him at a very rapid speed. He was about to run when the monster suddenly reversed and swam away. Chartrand said it was fifteen feet long with a humped back.

Adult

A year later, the Mugwump was featured in a regional newspaper. The mayor of Ontario called it "Old Tessie" and claimed it was four times the length of a man.

In 1982, another article ran in the *Temiskaming Speaker*. Kate Ardtree had heard stories of the Mugwump from her father. She had never seen it herself, but she said he once brought her a Mugwump scale the size of a dinner plate.

Ice fishermen Roger Lapointe and Dan Arney spotted a Mugwump while sitting in a fishing hut, also in 1982. They had cut a hole in the winter lake ice and dropped their lines in the frosty waters when things went haywire. The lines flew up in the air then fell back through the hole. Frightened, the men decided to head home. But as they walked back to the shore, they took one last look at the hole. A huge, black head was visible, its hungry eyes fixed on the men.

Dozens of eyewitness accounts of the Mugwump have been featured in the *Temiskaming Speaker* over the years. Do these stories prove the monster is real? No. But they do prove there is something mysterious to ponder in Ontario and Quebec.

Baby

NAHUELITO

(nah-hue-LEE-toe)

FIRST REPORTED	1897 OR 1922
LOCATION	ARGENTINA
CRYPTID TYPE	LAKE MONSTER
REALITY RATING	★ ★ ★

FACTOID: Nahuelito only reveals itself when the weather is warm and the wind is still.

EYEWITNESS ACCOUNTS: Nahuel Huapi Lake covers more than three hundred miles of ground at the foot of the Patagonian mountains in Argentina. And for centuries, local eyewitnesses have told the story of Nahuelito swimming in its vast waters.

The monster has been described as a huge water snake with humps, fish-like fins, and the neck of a swan. Some say it's fifteen feet long. Others say it's ten times that length. Regardless of its size, Nahuelito is said to be strong enough to overturn boats.

It rises to the surface of the lake, creating a sudden swell of waves and a spray of mist in the air. In the middle of the eighteenth century, when Argentina was under the rule of

Adult

King Charles IV of Spain, the stories were so concerning that the king ordered his men to launch an expedition to find Nahuelito. They failed.

Skull

In 1910, Canadian businessman George Garrett saw the monster while boating near the shore. He said the creature was between sixteen and twenty-three feet long, with a long neck that rose a full six feet above the surface. The *Toronto Globe* ran an article about Garrett's experience twelve years.

American Martin Sheffield saw the monster when he was searching for gold in Argentina that same year. Sheffield described a plesiosaur-like body with a long swan-like neck—a description that had appeared in Sir Arthur Conan Doyle's famous book, *The Lost World*, ten years earlier.

When Dr. Clementi Onelli, the director of the Buenos Aires Eco-Park, heard the eyewitness stories, he mounted an expedition too. Again, it was unsuccessful, but journalist Leonard Matters wrote about the search in the July 1922 issue of *Scientific American*.

The Onelli search won Naheulito the affection of many Argentine citizens. They petitioned the minister of the interior to refuse future permits for exploration to protect the monster. They also hoped to ban the use of dynamite and elephant rifles near the lake.

Some wondered if the Loch Ness Monster had crossed the North Atlantic—eight thousand miles—to make its way to Argentina. Some said nuclear power had caused the rise of Nahuelito. Both theories were rejected.

What is the truth? It's hard to say. But humanity will keep searching—no matter how long it might take to find out.

Baby

OARFISH

(OR-fish)

FIRST REPORTED	**1539**
LOCATION	**WARM TROPICAL SEAS AND OCEANS WORLDWIDE**
CRYPTID TYPE	**SEA MONSTER**
REALITY RATING	★ ★ ★ ★

FACTOID: There are three different species of oarfish that live deep in temperate and tropical bodies of salt water. *Regalecus glesne* is the most common of the three.

EYEWITNESS ACCOUNT: When Olaus Magnus created a map he called the Carta Marina—Map of the Sea—in 1539, it was considered the most accurate guide to the earth's great oceans. Scattered across the map were drawings of sea monsters— creatures seen by seafaring adventurers since the start of written history.

One of those monsters was over fifty feet long and weighed more than four hundred pounds. It had a horse-like face, large eyes, and a small toothless mouth. Its bony body was silver in

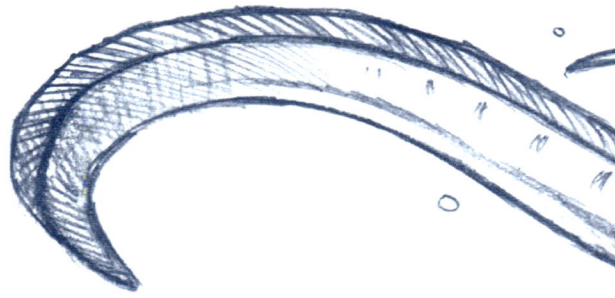

color with a red dorsal fin that ran the entire length of its spine. Was that gigantic monster real?

The answer is yes! The sea monster captured on the Magnus map was an oarfish. Regalecus glesne is its scientific name, but it's also known as a ribbon fish. And while it is rarely seen due to its habitat being thousands of feet below the ocean's surface, it has been found near death on or near beaches.

In October of 2013, snorkeler Jasmine Santana was startled when she came across an oarfish, complete with an eye the size of a half dollar, in the waters off Catalina

Adult

Island in Southern California. It was already dead when Santana found it, so she pulled it to the shallows for a closer look. It took more than a dozen people to carry the eighteen-foot fish to dry land.

Santana was a marine science instructor at the Catalina Island Marine Institute, so she quickly summoned her team. They called it a once-in-a-lifetime opportunity and learned that the rare creature had died of natural causes.

In recent years, scientists have proven that oarfish swim vertically, their heads pointing upward, their tails pointing toward the ocean floor. Marine biologists caught the healthy

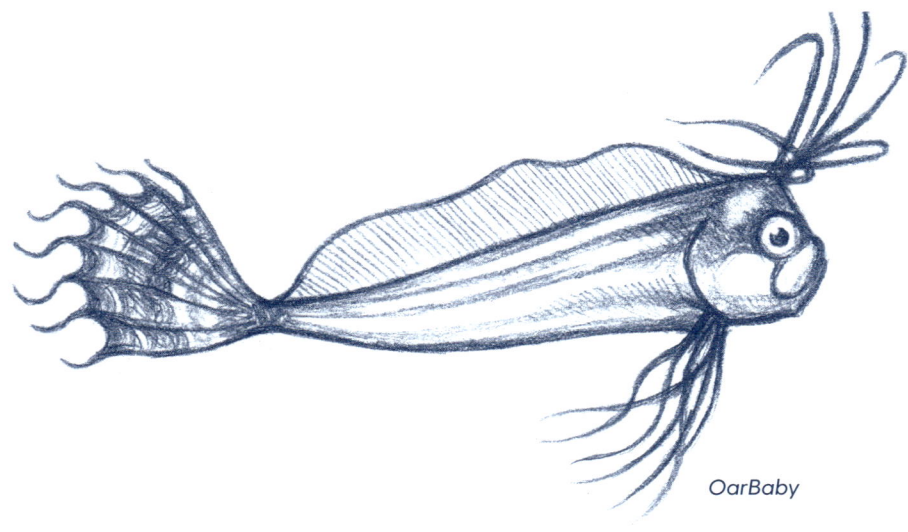

OarBaby

oarfish on film in the Gulf of Mexico from a small submarine. It used its red dorsal fin to stabilize its course. When it wanted to move at a faster pace, it undulated like a swimming snake.

It may be surprising to know the massive fish live on a diet of plankton, as do other giants of the sea. Some say oarfish can predict the coming of earthquakes or tsunamis due to their long, ribbon-like anatomy, but it's never been proven.

Since oarfish only come in contact with people when they are stressed—sick or dying they are seldom seen. Perhaps this is good, for as long as they stay hidden, they stay safe.

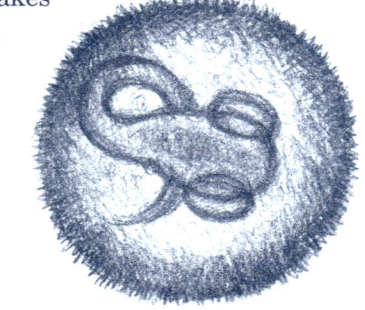

Egg

OGOPOGO

(OH-go-POE-go)

FIRST REPORTED	**1809**
LOCATION	**BRITISH COLUMBIA, CANADA**
CRYPTID TYPE	**LAKE MONSTER**
REALITY RATING	★ ★ ★ ★

FACTOID: A fifteen-foot statue of Ogopogo in Kelowna, British Columbia, gives Canadian children the chance to climb on its multi-humped back—in a make-believe way.

EYEWITNESS ACCOUNT: For decades, the legendary Ogopogo has been a staple in the city of Kelowna. Hundreds of eyewitnesses have filed reports of seeing a greenish-brown humped giant with the head of a horse gliding across Okanagan Lake. But

Baby

the stories didn't actually begin from sightings of a monster, they were started by early colonists in response to the indigenous people's belief in a lake spirit.

Before Europeans colonized North America, Okanagan Lake was a sacred space for the Syilx Okanagan. What is now known as Ogopogo was once called *n ʼx̌aʼx̌aitk*ʷ (pronounced "n-ha-ha-IT-koo"-koo"), and it was the sacred spirit of the lake, charged with protecting the people and animals that called the valley home. The Syilx Okanagan describe this spirit as dark in color with the head of a horse and the antlers of a deer. It is largely unseen but can manifest itself in the lake when it needs to be seen.

The Syilx Okanagan people would sometimes thank n ʼx̌aʼx̌aitkʷ with gifts of tobacco and sage, or salmon meat, when the catch was good.

When European fur traders arrived in 1809, they co-existed with the Syilx Okanagan people who had cherished the land for centuries.

Christian settlers were alarmed by the offerings and soon spread false rumors that the Syilx Okanagan lake spirit required blood sacrifice to guarantee safe passage across the lake. Stories of a vengeful serpent soon took hold, even if they were lies.

Few understand that Ogopogo is the misappropriation of a First Nations ritual of gratefulness, but the local Okanagan

Skull

Heritage Museum is working to change that, weaving Indigenous traditions into more recent stories, thanks to Westbank First Nation historians.

Moccasin Trails, an Indigenous tour company, offers paddling trips on Okanagan Lake where it shares the original traditions and reminds the visitors *n 'x̌ax̌aitkʷ* is a spiritual guide and protector, not a monster. Today, the Ogopogo is a playful Kelowna staple. T-shirts and stuffed animals of the cartoon-ish monster can be found in many local shops. Even the local hockey team has adopted the Ogopogo as their mascot.

Tourism officials adopted the name from an English folk song and offered a $1 million reward for proof of the creature's existence in the 1980s. In response, Greenpeace, an environmental watchdog, named Ogopogo an endangered species worthy of protection.

Do people still see Ogopogo in the water? They do. And, as long as evidence is being gathered, Ogopogo just might be real.

Adult

OKOBOJI

(oh-koe-BOW-gee)

FIRST REPORTED	1900s
LOCATION	IOWA, UNITED STATES
CRYPTID TYPE	LAKE MONSTER
REALITY RATING	★ ★ ★

FACTOID: Okoboji isn't the only strange marine animal sighted in Iowa. Six-foot-long paddlefish that weighed over two hundred pounds were once common there.

EYEWITNESS ACCOUNTS: Lake Okoboji in Iowa is said to have a massive monster lurking beneath its waves. Known as Okoboji after the lake of its origin, it has never eaten a swimmer or fed on livestock. It's done nothing to terrorize lake visitors, apart from coming too close to boats on the water.

Baby

The *Vindicator and Republican* in Estherville featured the story of a near miss in the early 1900s. The Bartletts, a married couple, were enjoying a day on the lake when

Skull

they noticed a commotion under the surface. They weren't sure what was causing the churning, but it was close to their boat and moving quite rapidly. It might have been a sea serpent, they said—or an ordinary fish grown to an extraordinary size.

Mr. Bartlett grew more and more fearful, thinking the enormous creature might shatter their boat and endanger his wife. He had no defenses. Suddenly, the chop in the water ended as the animal vanished. The couple didn't pretend to know what was causing the wake, only that it was large and powerful. They said they would never have believed it possible if they hadn't seen it with their own eyes.

Other eyewitnesses said the monster was 150 feet long from snout to tail. It was greenish- or bluish-gray with scales that made a repeating pattern. The stories continued into the 2000s, when a father brought his daughter to the lake for an ice cream cone and to watch the sun set.

When his daughter pointed to something thrashing out in the lake, the father saw a large, serpentine creature jump out of the water then splash down in an arcing motion. They saw a creature that was greenish-gray in color, with the head of a snake and a long muscular neck. When it submerged, they saw its stubby tail follow its body into the depths.

They thought it was gone, until it popped back out of the water and headed for the pier where they were standing. As it dipped under the pier, they felt the wood rock and rattle until the monster disappeared for the night.

Was it really a lake monster? Was it a huge fish? Perhaps time will tell.

Adult

RAYSTOWN RAY

FIRST REPORTED	1962
LOCATION	**PENNSYLVANIA, UNITED STATES**
CRYPTID TYPE	**LAKE MONSTER**
REALITY RATING	★ ★

FACTOID: Cryptozoologist Loren Coleman—the director of the International Cryptozoology Museum in Maine—thinks it's unlikely a monster could live in a lake created by human beings.

EYEWITNESS ACCOUNTS: The first sighting of Pennsylvania's Raystown Ray came in 1962, in the old Raystown Dam, though the details have been lost.

Constructed in 1905, the dam was removed in 1971 to make way for the Raystown Lake. Just 185 feet at its deepest point, it became the preferred spot for the annual Raystown Ski Club Water Show. That show

Skull

Adult

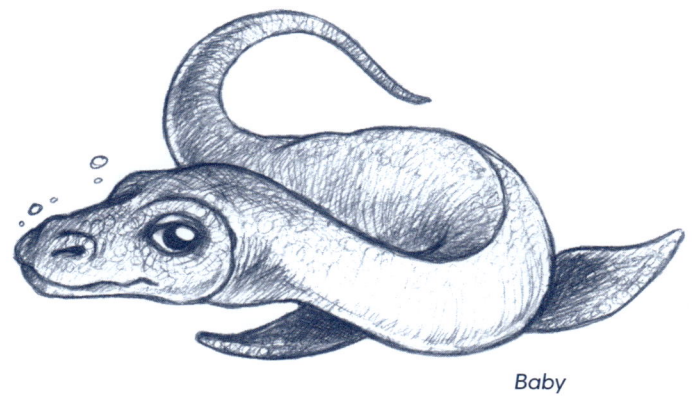

Baby

was almost canceled when the Raystown Ray was spotted near the ski ramps.

A new sighting came in 2007, when a tourist visiting Raystown Lake captured a very strange picture. It was a reptilian creature thirty feet long with a body that remained underwater, even as the long neck and head rose six to eight feet above the surface.

A television crew from the Syfy channel's *Fact or Faked: Paranormal Files* went to Huntingdon County in 2010 to investigate the mysterious sighting in the 8,300-acre man-made lake. As the cameras rolled, they took sonar readings from their boat, scuba dived through the murky waters at night, and photographed a floating log, hoping to replicate the tourist photo. They also towed a dead carp (fish) through the lake as bait, hoping to attract a hungry Raystown Ray.

In the end, *Fact or Faked* decided there was credible evidence for "something" living in the lake but proclaimed the mystery unexplained. The community hopes to change that status. Local stores sell patches, postcards, and T-shirts to celebrate the Raystown Ray, whether it's real or not.

Adult

Skull

SELMA

(SELL-ma)

FIRST REPORTED	**1750**
LOCATION	**NORWAY**
CRYPTID TYPE	**LAKE MONSTER**
REALITY RATING	★ ★ ★

FACTOID: The fishing village of Seljord uses a bright red sea serpent as its official coat of arms.

EYEWITNESS ACCOUNTS: The Seljord monster, known as Selma, has a long history. The first written reports came in 1750 and they've continued as time has ticked forward. Seljord, a fishing village, has deep roots in the sea.

Norwegian oceanographer Jan Sundberg set out to find Selma with help from Norwegian experts in the summer of 1998. But the team found no hard evidence.

Sundberg did not give up. Instead, he collected dozens of eyewitness accounts. They described Selma as being black in color and very big—between ten and forty feet long. But Selma is not a serpent. It has a thick middle and flippers to propel it through the water. Its head is like a horse's with no ears, and it has huge black eyes.

His second expedition featured acoustic microphones to survey the waters for Selma sounds. This time the scientist struck gold. The microphones picked up the sounds of an unknown mammal living in the fjords.

Encouraged, the scientists mounted a third expedition in 2000—the largest search yet. They brought sonar to probe the depths of the water. It was a great way to detect mass and motion in murky water, using sound waves to create a visual image.

Hoping to catch a baby Selma, the team also set a trap—an eighteen-foot-long cage meant to snare the little monster without causing harm. If the trap was successful, experts could collect DNA samples and set the mysterious animal free.

The trap did not work, but the sonar gathered some interesting sound readings. Analyzed in 2003 by researchers at the University of Copenhagen, it was determined that the vocalizations were made by a large unknown mammal.

In 2004, researchers allegedly captured a photo of a baby Selma, but experts have not yet confirmed its authenticity. So the search continues.

Baby

SHARLIE

(SHAR-lee)

FIRST REPORTED	**1920**
LOCATION	**IDAHO, UNITED STATES**
CRYPTID TYPE	**LAKE MONSTER**
REALITY RATING	★ ★

FACTOID: When fictional giant Paul Bunyan caught a huge sturgeon, his ox Babe launched it into the air. It fell into Payette Lake, where it's known as Sharlie today.

EYEWITNESS ACCOUNTS: Payette Lake in McCall, Idaho was formed by the deep cuts of a sliding glacier. As a result, it is eight miles wide and almost four hundred feet deep.

Railroad workers saw something strange in the lake in 1920. At first, they thought it was a log drifting in the currents. But when the huge dark shape started to move in an undulating motion, they realized it was no log. It was a mystery animal—a monster, they thought. It vanished as quickly as it appeared.

Adult

But it was seen again in 1944. The witnesses spotted a dark-colored animal about thirty-five feet long with a reptilian head, humps like a camel, and scaly skin.

The 1944 story was covered by *Time*. They called the creature "Slimy Slim" and documented thirty different eyewitness stories to confirm the description. Thomas Rogers, a businessman from Boise, Idaho, was one of them. He said the lake serpent swam past his boat just fifty feet away, moving roughly five miles an hour. Its raised head resembled a "snub-nosed crocodile" and its body was more than thirty feet long.

Curious people rushed to Payette Lake in search of Slimy Slim. Photographers camped out on the shores hoping to capture a picture to prove it was real. When the monster became popular, city leaders wanted to give it a nicer name. McCall's newspaper, *Star-News*, held a contest in 1954 and a winner was selected. "Sharlie" replaced "Slimy Slim."

Skull

Most locals believe Sharlie is a massive sturgeon. The popular game fish can grow to be more than eight feet long and weigh as much as three hundred pounds. They can also live fifty years.

Sharlie believers say waves that suddenly break in the middle of the lake are proof of the monster. Scientists say the waves are disturbances caused by a rapid change in barometric pressure.

The debate continues, but Sharlie is real to its fans in the city of McCall. It is the centerpiece of the city's annual winter carnival parade, where ice sculptors pay tribute each year.

In 2022, when paleontologists found *Serpentisuchops*, a twenty-three-foot-long fossilized plesiosaur in Wyoming, some thought it sounded like Sharlie. But the chance of Sharlie being a plesiosaur is slim. That doesn't stop Sharlie fans from hoping it's real and a distant cousin to Wyoming's marine reptile.

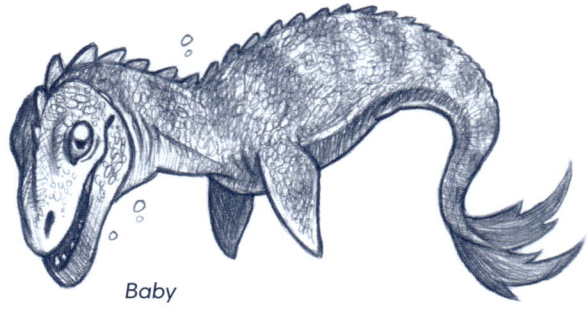

Baby

SOUTH SIDE SEA MONSTER

FIRST REPORTED	1976
LOCATION	NOVA SCOTIA, CANADA
CRYPTID TYPE	SEA MONSTER
REALITY RATING	★ ★ ★ ★

FACTOID: Fisherman Eisner Penney thought he was seeing a sunfish. But after watching the monster for twenty minutes, he knew it was something unknown.

EYEWITNESS ACCOUNTS: Penney was working the North Atlantic off the coast of Nova Scotia on July 5, 1976. After eight hours of fishing for haddock, cod, and halibut, he headed home.

It seemed like an ordinary trip to shore until he noticed something was following his boat—something big. The head of what Penney called a monster rose fifteen feet out of the

Skull

Adult

water and chased him for about five miles, gaining speed along the way. Just as Penney thought the creature would catch his boat and end his life, the monster vanished, leaving the fisherman rattled but grateful to be alive.

Penney didn't warn fellow fishermen Rodney Ross and his father about the monster when he got back into town. So they had no fear when they took their boat out two days later. But that would soon change.

The catch was good, so the father left Rodney on deck while he went below for a quick snack. That was when the silence descended. Even the seagulls disappeared.

MONSTER TURTLE OF THE BIG BLUE POND

FIRST REPORTED	**1981**
LOCATION	**IOWA, UNITED STATES**
CRYPTID TYPE	**LAKE MONSTER**
REALITY RATING	★ ★ ★

FACTOID: The prehistoric Archelon is the biggest turtle that ever roamed the earth. It grew to be fifteen feet long—the size of a compact automobile.

EYEWITNESS ACCOUNT: Big Blue is a pond that was created in Mason City, Iowa, in the early 1970s. When a brick and tile company closed, its fourteen-acre clay pit was transformed into a pond thirty-five feet deep. Stocked with rainbow trout, bass, bluegill, and northern pike, it's a great place to fish. And scuba divers train in the pond for deeper dives.

Close to playgrounds, the pond is a popular place for families. But reports of a Monster Turtle lurking in the thirty-five-foot depths of Big Blue pond have caused some concern.

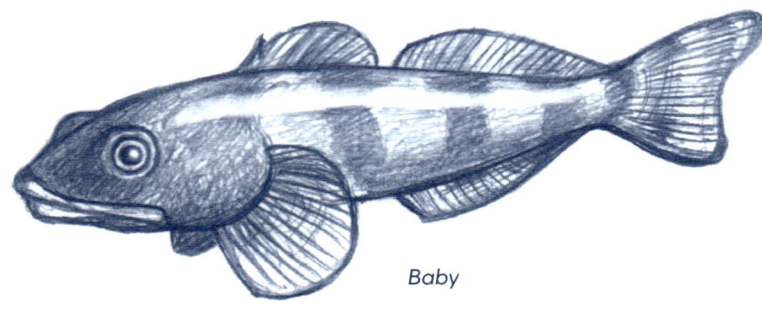

Baby

As Rodney noticed the eerie quiet, an odd sound began—a swishing. The fog was so thick, Rodney couldn't see what was making the noise until it was almost too late. A huge black mass came rushing through the water toward the boat—a mass with a hump and gigantic, black eyes.

Rodney's father tried to start the engine to try to escape when the monster rose up fifteen feet out of the water. It was covered in barnacles and had a huge mouth with rows and rows of dagger-like teeth. "It was after us," Rodney said. Just as they reached the dock, the monster rammed the boat and disappeared.

Fishermen Edgar Nickelson and his son Robert laughed when they heard Rodney's story. But as they set out to fish two days later, they were no longer laughing. The monster revealed itself to them too. They thought it was a whale, so Edgar turned on a noisemaker to scare it away. It ignored the noise and headed straight for them. "If there is a devil," he said, "that was it."

For weeks, no one dared return to fishing for fear of the monster. Five men on three different boats had seen a sea monster, and they keep watch for its return, just in case.

STORSJÖODJURET/ STORSIE

(STOE-hure-YOUR-et) (STORE-see)

FIRST REPORTED	**1635**
LOCATION	**SWEDEN**
CRYPTID TYPE	**LAKE MONSTER**
REALITY RATING	★ ★ ★

FACTOID: Religious leader Morgens Pederson said Storsie was created by two trolls using a magic potion.

EYEWITNESS ACCOUNTS: Father Pederson may have been the first to see the Lake Storsjön monster in 1635, but he was not the last. A mechanic named Martin Olsson saw it in 1878 as he fished the lake on a clear, sunny day.

Olsson felt something was watching him. When he turned around, he spotted the monster's long, slender neck and its thick gray body glistening in the sunlight. It was as wide as a full-sized man. The monster's

Skull

Adult

snake-like head seemed too big for its neck. Olsson did not see scales, but the skin did seem fish-like. And its eyes captured his imagination—bright red and focused on him.

Startled, Olsson dropped his fishing pole and tried to move out of reach. If he moved too quickly, he feared the creature might attack. If he moved too slowly, it might think he was vulnerable. So he decided to row slowly and quietly. To his relief, Storsie lost interest and disappeared underwater.

Three centuries later, Storsie resurfaced. Fisheries officer Ragnar Björks was checking fishing permits on the lake when he encountered Storsie. He estimated it was eighteen feet long when he saw its tail break the surface of the lake. When the

Baby

rest of its body appeared, Björks was terrified and felt the need to defend himself.

When it swam close, Björks hit it with his oar. He was hoping to scare Storsie away, but it was not frightened. It was angry. It slapped its tail against the surface of the lake so hard, it lifted Björk's boat a full twelve feet in the air before it vanished again. Björks had been a nonbeliever when it came to Storsjöodjuret, the monster of Lake Storsjön—until it slapped a little sense into him that day.

Hundreds of other people have made similar claims. The stories do vary, but most say Storsie is huge with a long neck, a humped back, and a lengthy tail. The upper body is a gray-brown color and its belly is yellow.

Proof has yet to arrive, but for almost twenty years, Storsie was protected by Swedish law. That legislation is no longer enforced, but if Storsie reappears, lawmakers may offer new protections. Time will tell.

TAHOE TESSIE

(TAH-hoe TESS-ee)

FIRST REPORTED	1848
LOCATION	NEVADA/CALIFORNIA, UNITED STATES
CRYPTID TYPE	LAKE MONSTER
REALITY RATING	★ ★ ★

FACTOID: The Washoe people of Nevada and California have stories about Water Babies that live at the bottom of Lake Tahoe. Some believe they warn of a coming death. Others believe they foretell blessings.

EYEWITNESS ACCOUNT: Tessie sightings began during the California gold rush hit in 1848 when hundreds of new people came to seek their fortunes.

I. C. Coggin, a San Francisco socialite and musician, wrote about his Tessie sighting in the *San Francisco Examiner* in 1897. He described a serpent-like monster six hundred feet

Skull

long with a head fourteen feet wide and huge jet-black eyes. It slithered by, crushing forest brush, as Coggin hid behind a tree. He knew people would not believe his story, but he was sure it would eventually show itself to other witnesses and prove he wasn't a liar.

Coggin was right. Many more eyewitnesses did come forward—so many that famed French oceanographer Jacques Cousteau joined in the search for more evidence. In the 1970s, Cousteau piloted a small submersible submarine in Lake Tahoe. When asked what he discovered, Cousteau said "The world isn't ready for what's down there."

Reporters at the *San Francisco Chronicle* wrote about another sighting on July 12, 1984. Tahoe residents Patsy McKay and Diane Stavarakas were hiking above the lake when they spotted something odd in the water. They said it was seventeen feet long with humps between its head and the tail. McKay said it surfaced three times and left a wake each time.

Most local scientists believe Tessie is a large fish—perhaps a sturgeon—not a monster. Believers disagree. Mississippi

Adult

resident Ashutea Young gave those believers hope when she visited Lake Tahoe in 2021 and captured what she thinks was Tessie video footage. Did that prove Tessie was real? Not yet. But the lake monster's fans aren't ready to give up hope.

Baby

TRUNKO

(TRUN-koe)

FIRST REPORTED	**1925**
LOCATION	**SOUTH AFRICA**
CRYPTID TYPE	**SEA MONSTER**
REALITY RATING	★ ★

FACTOID: The man who discovered this creature did not call it Trunko. He said it looked like a huge polar bear. The name Trunko came from cryptozoologist Karl Shuker.

EYEWITNESS ACCOUNT: Weird stories were popular in 1920s newspapers. That's how the *Call-Leader* in Elwood, Indiana, came to feature Trunko, an exotic monster from South Africa.

According to the article, Hugh Ballance was strolling on the beach in October of 1924 when he saw two whales doing battle with a mysterious creature just a few yards away. Ballance grabbed his glasses to get a better look and saw an enormous,

Skull

polar-bear-like creature. He said the monster used its tail to kill both whales then collapsed on the shore.

The exhausted monster was forty-seven-feet long, ten-feet wide, and five-feet high. Ballance saw no head, but there was a lobster-like tail (ten feet long) on one end and a five-foot trunk on the other. At the mouth of the trunk, Ballance saw a pig's snout. Snow-white hair, ten inches long, covered its entire body.

The monster lay motionless for ten days, according to Ballance, then it crawled into the waves and swam away, heading toward the southeast. Ballance never saw blood on the creature, so he assumed it had no blood. But it had vanished before scientists could examine the body.

Baby

Adult

An article ran in the London *Daily Mail* in December of 1924 with the headline, "Fish Like a Polar Bear." The story later made its way to Indiana.

When a beast covered in long white hair washed up on a beach on Dinagat Island in the Philippines in 2017, people wondered if it was another Trunko. But experts determined the beast was likely a decomposing whale.

Was Trunko also a dead whale? Most experts say it was. But without bodies to analyze, it's impossible to say for sure.

Weird Whales

In March of 2023, scientists observed strange behavior from a pair of whales. Instead of skimming the water to feed on tiny krill, they hovered at the top of the water, mouths opened wide, and waited for a school of fish to swim in.

Science calls it trap feeding or tread-water feeding. But modern man wasn't the first to witness the strange practice. Thirteenth-century Norsemen called the hungry beasts hafgufas. And they sometimes thought it was a sign of the Kraken or Merfolk.

One map created in 1658 suggested whales were the size of small islands. One witness said, "The greatest of whales could not chase fish, but caught them through cunning." As they prepared to feed, the whales let out enormous burps then gulped up fish. At first, experts thought they were making room for a meal by burping. Today, scientists think the belches release partially digested fish in the whale's stomach as bait—to lure more fish to eat. Ancient reports said the burps smelled like rotten cabbage.

Trap feeding was confirmed as a real feeding practice when experts spotted whales using the technique in 2011 and again in 2023.

UB-28 SEA MONSTER

FIRST REPORTED	1915
LOCATION	IRELAND
CRYPTID TYPE	SEA MONSTER
REALITY RATING	★ ★ ★

FACTOID: During World War I, German submarines called U-boats sank ships carrying supplies for their enemies.

EYEWITNESS ACCOUNT: In July 1915, as World War I raged on in Europe, German UB-28 submarine Commander Georg-Gunther von Forstner set his sights on the British steam ship *Iberian* sailing through the North Atlantic near Ireland. He ordered his torpedoes fired and they hit the unsuspecting ship square in its center.

The *Iberian* sank so quickly, the Germans were amazed. Twenty-five seconds after it vanished into the depths of the ocean, a massive explosion sent wreckage flying into the air as von Forstner and six of his key officers watched. Most of the debris was predictable. The sea monster was not.

Adult

Von Forstner recorded the event in his commander's logbook, describing a gigantic aquatic animal launched eighty feet into the air.

"We were unable to identify the creature," he wrote, "but all of us agreed that it resembled an aquatic crocodile, which was about sixty feet long with four limbs resembling large webbed feet, a long pointed tail, and a head which also tapered to a point."

The creature went under the water ten to fifteen seconds after it was spotted, so von Forstner was unable to take a photograph.

None of the witnesses are alive to confirm the story today. But the fact that von Forstner made mention of the event in his official log gives the story credibility. Making up such a story would have damaged the commander's reputation as a serious military man. So the odds are good he and his officers were being honest.

Baby

Most ocean lifeforms flee the sounds and dangers of war, so it seems unlikely the creature was swimming near the *Iberian* when it blew up. Some wonder if the *Iberian* had the massive mystery animal on board for transport when it blew up. But that is just speculation.

Von Forstner's account remained buried in history without notice until 1933, when stories of the Loch Ness Monster hit newspapers across Europe. Von Forstner wrote an article about the Scottish mystery for a German newspaper, sharing his own sea monster account with a suddenly curious readership.

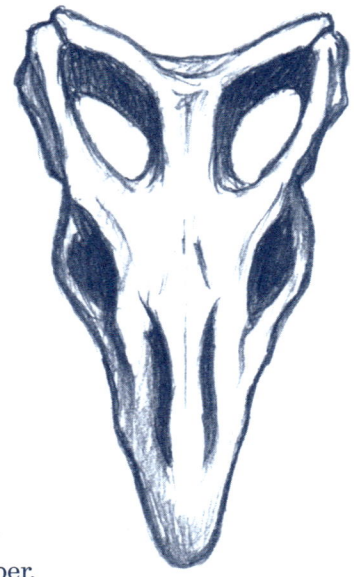

Skull

Did the crew of a UB-28 German submarine really see a mysterious sea monster in 1915? Famed cryptozoologist Bernard Heuvelman said yes. He called the monster a "marine saurian" and said it likely escaped extinction in a prehistoric era.

It's impossible to know for sure, but von Forstner's epic tale, critically recorded in an official military logbook, certainly makes it seem entirely possible.

WALLY

(WALL-ee)

FIRST REPORTED	1885
LOCATION	OREGON, UNITED STATES
CRYPTID TYPE	LAKE MONSTER
REALITY RATING	★ ★ ★

FACTOID: Some people credit the first Wally stories to the Nez Percé. But confirmation has been hard to find.

EYEWITNESS ACCOUNT: The unconfirmed Nez Percé legends speak of two lovers paddling across Wallowa Lake, only to be dragged to their deaths by a fearsome monster. But the first documented stories appeared in the *Wallowa County Chieftain* on November 5, 1885.

A prospector who preferred to remain anonymous was rowing across the lake in a small boat. Just as the sun started to set, he saw something strange. It looked like an animal about fifty yards from the skiff.

Skull

Adult

The mysterious animal's head and neck were ten to twelve feet out of the water. When it noticed the prospector, it submerged to hide. Astonished, the man sat in silence, wondering what he had just seen. When the creature reappeared, it sounded a mournful cry.

Baby

The prospector saw a large, flat head like a hippopotamus and a ten-foot-long neck as thick as a man's body. As it glided away from his rowboat, the prospector watched until it grew too dark.

The next well-documented case came in 1978, when a couple saw the lake monster that came to be known as Wally twice. The first sighting revealed the monster had three humps that broke the surface of the lake. The second time they saw Wally, they said it appeared to be a snake-like creature twenty feet long.

Bert Repplinger and Joe Babic saw Wally three years later in Lake Wallowa. They claimed Wally had a long neck topped with a three-foot-long head.

When divers from the Blue Mountain Divers' SCUBA team searched the lake in June of 2019, they found no signs of Wally. But they did find barrels of toxic waste and pipe bombs.

Perhaps Wally will reveal himself the next time a crew of divers mounts a search.

Adult

WHITE RIVER MONSTER

FIRST REPORTED	1915
LOCATION	ARKANSAS, UNITED STATES
CRYPTID TYPE	RIVER MONSTER
REALITY RATING	★ ★ ★ ★

FACTOID: Arkansas passed legislation to give the White River Monster protected status. It is illegal to harm the mysterious creature.

EYEWITNESS ACCOUNT: Historians believe the first sighting of the White River Monster was in 1915, but Ethel Smith didn't come forward with the details for another fourteen years. In 1924, she admitted she and her husband heard a loud "blowing noise" coming from the river. They looked for the source of the noise and saw a huge gray creature covered in a crusty coating. It frightened the woman so much she didn't tell her story until she saw farmer Bramlett Bateman's story in the local newspaper in 1937.

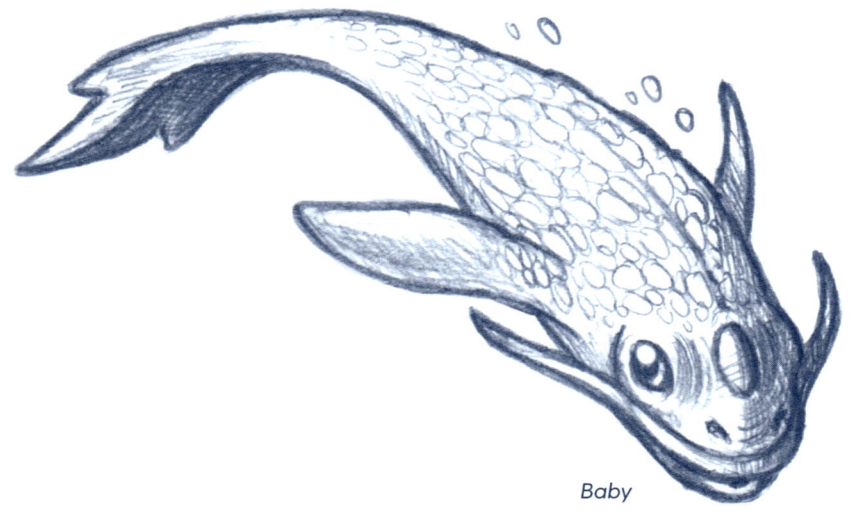

Baby

Bateman saw the monster in July of that year. He said it was about twelve feet long and four feet wide. He was so sure of what he'd seen that he signed a sworn statement to prove he was telling the truth. Three other people came forward with similar stories, including a deputy sheriff.

A plan was hatched to capture the monster with a giant net. As it was being constructed of sturdy rope, divers searched the location where Bateman had seen the creature but found nothing. Eventually, the plan was scrapped when the capture team ran out of money.

In 1971, another witness saw a gray creature with peeling skin and a horn at the center of its forehead swimming in the river. It was twenty feet long with sharp fins along its spine.

A trail of huge fourteen-inch, three-toed footprints added weight to the story.

In 1980, cryptozoologist Roy Mackal announced that the White River Monster was an elephant seal. Joe Nickell, a writer at the *Skeptical Inquirer*, said Mackal was close with that theory. But he believes a hooded seal is a better fit. They are common in Florida, which is only a few hundred miles from Arkansas.

Other experts say a manatee would be a more likely candidate. They often travel from the sea to freshwater rivers. That would make the White River Monster real but relatively ordinary—not mysterious. More information is needed to be sure of any theory. So, the search continues.

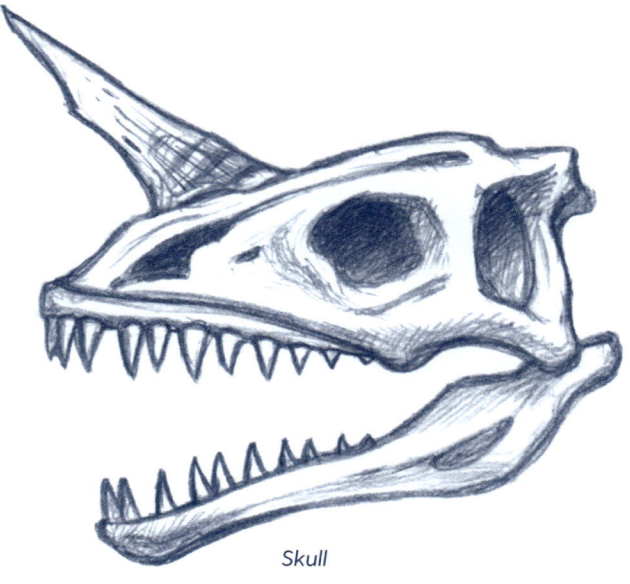

Skull

YACUMAMA

(YAK-oo-ma-ma)

FIRST REPORTED	**1900s**
LOCATION	**PERU**
CRYPTID TYPE	**RIVER MONSTER**
REALITY RATING	★ ★ ★ ★

FACTOID: British explorer Percy Harrison Fawcett was born in 1867. He spent decades searching South America for amazements, including the Yacumama, "the mother of water."

EYEWITNESS ACCOUNT: Fawcett first encountered Yacumama on the banks of the Amazon River. As his team searched the region's tropical rainforests, they spotted an enormous serpent

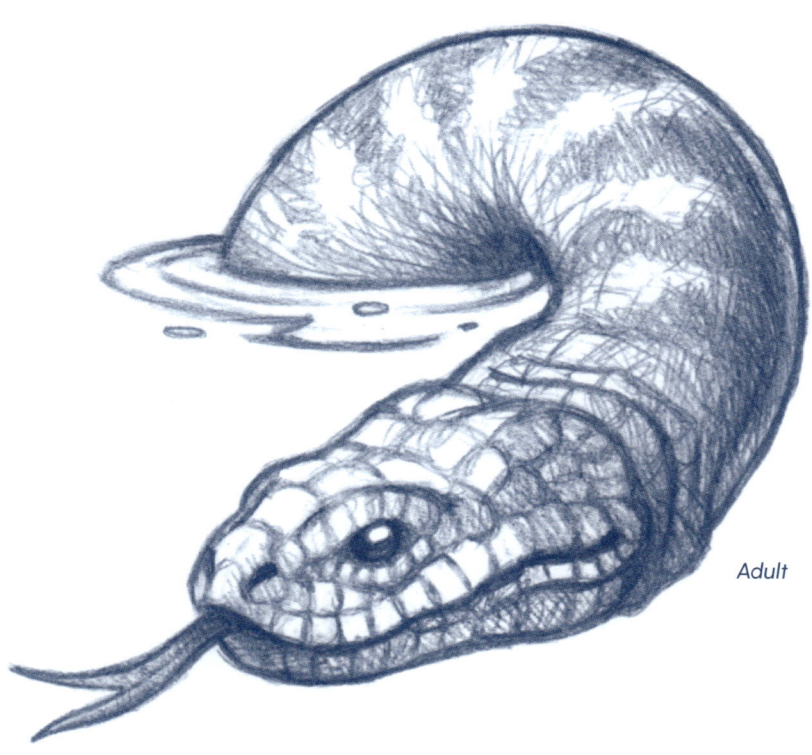

Adult

sixty feet long—forty-five feet on dry land, another fifteen feet in the river. Fawcett knew Peru's Indigenous peoples were aware of large anacondas in the rainforests—dangerous snakes they avoided. But he thought this sixty-two-foot monster might dwarf the snakes described by the locals.

The stories were not new. For centuries, the people of Peru and three other South American countries—Paraguay, Argentina, and Brazil—spoke of the Yacumama, a god that protected the rainforest and the rivers. Those who did not respect nature had little chance of escaping the jaws of the powerful snake.

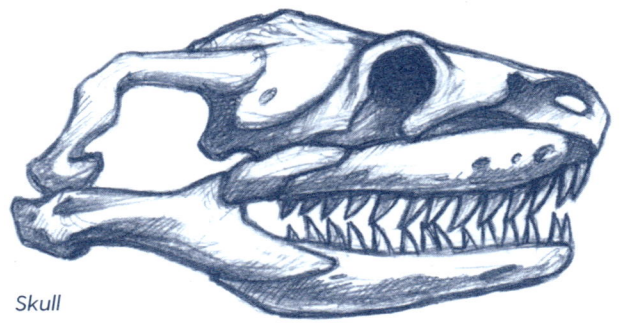

Skull

When it was on the hunt, the Yacumama fired powerful currents of water from its mouth to take down prey. Fishermen and their boats have allegedly vanished without a trace—consumed by Yacumama, the guardian of the Amazon.

In the 1900s, two men set out to kill the Yacumama. They loaded their boat with explosives and detonated them, hoping to end the dangers. Yacumama allegedly rose from the river's depths, covered in blood, but it did not die. It swam away as the men hid. They feared Yacumama would seek revenge.

The river monster may be a giant anaconda. But some say it could be a *Titanoboa*—a prehistoric giant that lived millions of years ago. Believed to have been fifty feet long and three feet wide, the snake was the rainforest's top predator after dinosaurs vanished sixty-five million years ago.

Sightings have been common throughout the years, but Irish lithographer Mike Warner used a new tool to search in 2009—satellite imagery. Warner found visual evidence that suggests Yacumama may still exist.

Warner thinks his satellite imagery discovered a gigantic caecilian—a snake-like amphibian common to South and Central America, Africa, and southern Asia. More than two hundred species of caecilians exist today, but most are only eighteen to twenty-two inches long. A five-foot-long giant known as *Caecilia thompsoni* was found in Colombia. And Mike believes more oversized individuals could be out there. He'll keep searching for proof. Until that happens, Yacumama will remain a mystery.

Baby

Appendices

RESOURCES

Sea Monster Movies

The Little Mermaid, 2023, rated PG

The Sea Beast, 2022, rated PG

Luca, 2021, rated PG

Pacific Rim Uprising, 2018, rated, PG-13

Song of the Sea, 2014, rated PG

Percy Jackson: Sea of Monsters, 2013, rated PG

Scooby Doo: Curse of the Lake Monster, 2010, rated PG

Ponyo, 2008, rated G

The Water Horse: Legend of the Deep, 2007, rated PG

The Little Mermaid, 1989, rated G

Creature from the Black Lagoon, 1954, rated G

20,000 Leagues Under the Sea, 1954, rated G

Sea Monster Video Games

What Lives Below, Coming Soon

Earth Atlantis Elite Edition, 2022

Fantasy Friends: Under the Sea, 2021

Oceanhorn: Monster of Uncharted Seas, 2015

Sea Monsters: A Prehistoric Adventure, 2007

Sea Monster Fiction Books

The Pearl Hunter, by Miya T. Beck, HarperCollins, 2023. Ages eight and up

The Monster Mission, by Laura Martin, HarperCollins, 2021. Ages nine and up

Madre de Aguas of Cuba, *The Unicorn Rescue Society* Book #5, by Adam Gidwitz, Penguin Random House, 2020. Ages eight to ten

Malamander, The Legends of Eerie-on-Sea Book #1, by Thomas Taylor, Walker Books, 2019. Ages eight and up

The Water Horse, by Dick King-Smith, Yearling, 1990 (republished 2007). Ages six and up

GLOSSARY

ALLEGEDLY: Claimed without proof.

ANACONDA: A boa constrictor snake native to South America.

ANONYMOUS: Unnamed.

BLUE HOLE: A deep marine cave or cavern in the waters of tropical regions such as Jamaica.

CULTURAL MISAPPROPRIATION: Using something without permission or connection.

DNA: Distinctive building blocks possessed by each living thing on earth that define its type or species.

INDIGENOUS: The earliest inhabitants of a place. Also, occurring naturally in a given geographic region.

INHABIT: To live in or occupy a place or region.

MAGNITUDE: Great size.

MYTHOLOGY: Narratives about the early history of a people.

NUCLEAR: Related to the nucleus of an atom.

OCEANOGRAPHER: A scientist who studies the ocean and its life-forms.

PELT: The fur or skin of a dead animal.

PETROGLYPHS: Ancient human-made rock carvings.

SAMURAI: Ancient Japanese military men.

SERPENTINE: Snake-like.

SPECIES: A group of creatures with many physical traits in common.

SUPERNATURAL: Beyond the natural world.

TESTIMONY: The statement of fact, according to a witness

TOURISM: The professional attempt to draw visitors to a place

UNDULATE: To move with an up-and-down motion, not side to side.

UTERUS: An organ in female mammals for nourishing the young during development.

INDEX

Turkey, monster in. *See* Lake
Van Monster

U

UB-28 Sea Monster, 171–74, 172*fig.*
United States, monsters in
  Alkali Lake Monster, 2–5
  Altamaha-ha, 6–9
  Bear Lake Monster, 10–13
  Bessie, 17–21
  Cape Ann Sea Serpent, 26–29
  Champ, 30–33
  Chessie, 34–39
  Edizgigantus, 42–44
  Flathead Lake Monster, 45–48
  Gonakadet, 52–55
  Iliamna Lake Monster, 60–62
  Lake Chelan Dragon, 80–83
  Memphre, 102–5
  Mishipeshu, 111–14
  Monster of the Susquehanna,
    121–24
  Monster Turtle of the Big Blue
    Pond, 125–28
  Okoboji, 144–46
  Sharlie, 153–55
  Tahoe Tessie, 163–66
  Wally, 175–77
  White River Monster, 178–81
*Unnatural History: True Manitoba
  Mysteries* (Rutkowski), 101
US Fish and Wildlife Service, 39
US Grant, 45
Utah, United States. *See* Bear Lake
  Monster

V

Van Yüzüncü Yıl University, 84–86
Venus girdle, 51
Vermont, United States
  Champ, 30–33
  Memphre, 102–5
Victoria Daily Times, 42–43

Vincent, Dick, 100
Vindicator and Republican, 145

W

W. B. Eddy (ship), 32
Wade, Jeremy, 97
Wainman, Amy, 49, 51
Walgren Lake Monster. *See* Alkali
  Lake Monster
Wallowa County Chieftain, 175
Wally, 175–77, 176*fig.*
Ward, Edward, 35
Warner, Mike, 184
Warren, Jeff, 8
Wasgo. *See* Gonakadet
Washington, United States
  Edizgigantus, 42–44
  Lake Chelan Dragon, 80–83
Water Babies. *See* Tahoe Tessie
Water Horse, The (film), 69
water horse. *See* Kelpie
whales, 60, 70, 72, 167–68, 70. *See also*
  Trunko
White River Monster, 178–81, 178*fig.*
White, Bob, Sr., 36
Whyte, Constance, 92
Widder, Edie, 74–75
worldwide monsters. *See also*
  lake monsters; river monsters;
  worldwide monsters
  Merfolk, 107–10
  oarfish, 137–40

Y

Yacumama, 182–85, 182–83*fig.*
Young, Ashutea, 166
Yunus, prophet, 59

Z

Zardulu, artist, 9
Zonn, Froy, 120
Zulu, people, 63–64

About the Author

KELLY MILNER HALLS wrote her first cryptozoology book in 2006 and fell in love with the mysterious animal stories. *Cryptid Sea Monsters* is her fourth crypto-book and *Cryptid Babies*, written with her youngest adult child, Ness Halls, will be her fifth. She has also written books about dinosaurs, mummies, aliens, ghosts, the *Titanic*, and an even longer list of fascinating nonfiction subjects. To learn more about her, visit WondersofWeird.com, or email her at KellyMilner@aol.com.

About the Illustrator

RICK SPEARS was born in California in the third quarter of the twentieth century, relocating to Georgia five years later. As with most young children, he was fascinated by dinosaurs and marine reptiles and spent many hours drawing pictures of prehistoric animals in various scenes of combat. His dinosaur and cryptid art has found its way into various books, magazines, websites, and even a board game. His sculptures, including that of the river monster Altamaha-ha, can be seen in several museums and collections throughout the southeastern United States. Rick lives in Georgia with his family, who find his weird projects just a part of everyday life.